Chemical Thermodynamics and Information Theory

with Applications

Chemical Thermodynamics and Information Theory

with Applications

Daniel J. Graham

CRC Press
Taylor & Francis Group
Boca Raton London New York

CRC Press is an imprint of the
Taylor & Francis Group, an **informa** business

CRC Press
Taylor & Francis Group
6000 Broken Sound Parkway NW, Suite 300
Boca Raton, FL 33487-2742

First issued in paperback 2018

ISBN-13: 978-1-4398-2087-2 (hbk)
ISBN-13: 978-1-138-37427-0 (pbk)

Visit the Taylor & Francis Web site at
http://www.taylorandfrancis.com

and the CRC Press Web site at
http://www.crcpress.com

To Teresa

Contents

Preface

Chemistry majors receive thermodynamic instruction of one form or another during all four college years. Individuals who pursue graduate studies gain further exposure. The curriculum is layered and standard given decades of development. First-year students learn the basics of work, heat, state functions, and reaction spontaneity. The advanced topics delve deeper into the thermodynamic laws, equations of state, and phase transitions. The fundamentals and applications are inseparable throughout. A student cannot understand how household refrigerators work without examining the properties that determine inversion temperatures.

Yet, thermodynamics and chemistry appear to be at a juncture. The intersection of subjects has enjoyed the maturity of years, and a rich combination of theory and applications. The backdrop, however, is no longer confined to classrooms and stand-alone textbooks. This is because information has leaped to the forefront as a fundamental and, at the same time, practical resource. It carries the same significance as heat, work, and variables of state, being closely related to the entropy of a system. Information operates as a fuel for some processes while it is a distinct product of others. The advances in thermodynamics transpired largely during the industrial revolution. Much of chemistry today, however, is threaded by programming, computation, and data analysis. The author believes that the thermodynamic curriculum should reflect this better by expanding the scope of fundamentals and the application of elementary models. If energy is a system's capacity for performing work and transferring heat, then information is the capacity for controlling how the work is performed and heat is exchanged.

This book is not the first to intertwine information and a physical science such as chemistry. The author was introduced to the fundamentals of information by well-known texts, in particular ones by Morowitz and Brillouin. Further, the information aspects of biopolymers—proteins and polynucleotides—figure in several places, from high school classes to the popular press. Most recently, new roots of information have been established in electronic structure theory, quantum computation, and drug discovery, to name a few areas.

This text nonetheless aims at a few new things while certainly not trying to address everything. The goal is to provide a fresh perspective of select topics, such as state transformations, heat and work exchanges, and chemical reactions. These processes do not occur by themselves for a system, but rather in cooperation with the surroundings and with information as the programming currency. The treatment is quantitative to the extent that it employs basic calculus, probability, and statistics. Wherever possible, however, the more intuitive elements of information and thermodynamics have been emphasized. Further, the major ideas have been presented less via derivation and more by example. As a result, the material should be appropriate for intermediate students and beyond in special topics classes or for self-study. Just as important, it is hoped that new perspectives and exercises are provided for instructors who will relay them to their clientele.

This book has a number of sources. They include (1) three shelves of treatises on thermodynamics, statistics, and computation—these have been purchased, borrowed, and inherited over the years; (2) regular teaching assignments in general and physical chemistry; (3) enlightening consultations with colleagues and students; and (4) the author's curiosity. Most important, the author is appreciative to his own teachers for intellectual foundations and inspiration, in particular Walter Rudolph and Drs. Eric Hutchinson, Robert Pecora, Bruce Hudson, Suzanne Hudson, Tien-Sung Lin, Sam Weissman, Ron Lovett, Richard Clarke, and William R. Moore. The author is equally grateful to his Loyola University colleagues, especially Donald May, Bruno Jaselskis, Carl Moore, Sandra Helquist, Jacob Ciszek, and Gordon Ramsey who have contributed helpful comments and criticism. Last, the author is appreciative of Lance Wobus, Kathryn Younce, and Linda Leggio of Taylor & Francis/CRC Press for their expertise, advice, and editorial assistance.

1 A Qualitative Look at Information

Information is examined in qualitative terms. The examples are drawn from both the macroscopic and microscopic scale. These set the stage for quantitative aspects discussed in the chapters that follow. The suggested exercises address several contexts where information is central.

1.1 THE NET OF INFORMATION

The word *information* motivates much discussion. It prompts almost 3 billion Google hits. "Everything is information" in the view of the late John Archibald Wheeler [1]. "Information is physical" according to Rolf Landauer [2]. Information is power in the eyes of educators, politicians, and computer hackers. Information is removed from the universe by black holes as proposed by Stephen Hawking and disputed by Leonard Susskind [3,4]. Information changes the state of a recipient [4]. According to Carolyn Marvin, "information cannot be said to exist at all unless it has meaning, and meaning is only established in social relationships with cultural reference and value" [5]. Information is that which appears planned [6]. Information is anything that alters a probability distribution [7]. *Merriam-Webster's Unabridged Dictionary* lists seven headings for *information*. The first definition is "an endowing with form" [8]. Other pronouncements can be cited. Gregory Bateson defines information as a "difference that makes a difference; if there is no difference, there is no information" [9]. "The transmission of information is impossible save for a transmission of alternatives," according to Norbert Wiener [10]. Information of certain genres is decried in the song "Satisfaction" by The Rolling Stones [11]. The purpose of this paragraph is not to fuel controversy. It is to show that a familiar term casts a multicontextual, unusual net.

Nets have threads. One is that information is highly accommodating of digital representations. If one reads this paragraph, one internally generates a set of electrical signals. If the reading is accompanied by speaking, one sparks the brain electricity plus a collection of air pressure waves. The electric and acoustic signals vary with the individual and environment. Yet, the kernel of information is captured quite effectively via symbols printed on the paper. The information can be communicated to equal effect using chalkboards and computer screens. In each venue, the representation is obtained by combinations of letters of a finite alphabet.

Note the power of digitization. Every unit of the alphabet means the same thing no matter what font size or style is used. All of the following are representations of the letter c:

$$c \quad c \quad c \quad c \quad c \quad \mathbf{C} \quad c \quad C$$

The symbols B, d, 1, ? and so forth can be used to make the point just as well, namely, that the units are themselves not tunable. Rather, a message derives from the capacity of the source to string units together and of the receiver to register the combinations. There are 51 [i.e., $(1 \times 26) + (26 \times 1) - 1$] two-letter combinations that contain letter c drawn from the English alphabet. The number increases to 53 if the space unit is included as an option. Clearly, the number of distinguishable combinations grows exponentially with the cluster size and available symbols. There is no need to worry about exhausting the possibilities for license plates or the nuances that can be imbedded in alphanumeric messages.

A related thread is that information accommodates more than one representation. Beethoven imagined music in complex ways. Yet, he communicated his thoughts using the digital formats of black spots, horizontal and vertical lines, clef symbols, and so forth—the notation of Western music. And he always had more than one way to convey a message. The same musical note can be represented using any of the treble, bass, tenor, or alto clef systems. The assemblies of notes can be communicated to equal degree via different key signatures. It is apparent that one information format can be substituted for another. A vital property of information is that it is virtually always fungible.

The reliance on digital formats requires a structure dependence. Alphabet characters can be strung in various combinations, but not any arrangement will do. The letters e, a, and t, linked one way refer to food consumption. Combined another way, they bring to mind a late afternoon beverage. Still another arrangement evokes food consumption in the past tense. Structure is critical to the message, and thus the content and impact of information. The receiver of messages from a source can only comprehend the messages if the structure rules are well established in advance.

Another property of information is that it admits reproduction. Printed texts and drawings can be hand- and photocopied. Their digital versions can also be burned on compact disks and copied to flash drives. The reproduction of information was limited in the Dark Ages due to the scarcity of paper and literate humans. Matters are quite the opposite in this electronic age. The typical American consumes about 34 gigabytes of electronic information daily. This is an increase of about 350% over the past three decades [12]. Citizens do not live by food alone for there are the dietary requirements of information.

This points us to another property. Information is highly accommodating of transformations. The texts and drawings that can be copied just as easily can be erased in parts and appended in others. Information is important for all the reasons mentioned so far. That it is mutable should occupy a prominent place on the list. Note that the transformations need not be planned or orderly. Many mutations indeed transpire by accident. The alterations of genetic material and computer files come to mind [13]. The learning of a new word, phrase, or concept on the part of an individual marks the transformation of neurological information, in most cases irreversibly.

Transformations and irreversibility mean that there is an energy price tag attached to information, its production, and consumption. Beethoven was a source long ago. Today a reader of his scores must work to process the information. The same holds for a listener or a performer of the music. Information processing is not free of charge, even when dollars appear nowhere in the transaction.

Yet work is not the only currency. Heat also plays a role in all manners of information processing. Beethoven transferred thermal energy from his hand to a pen as he applied notes to the paper. Heat was generated and dissipated by the pen at the point of contact. Heat is dissipated within the piano keys or violin strings by their contact with a performer. Heat is dispersed in the listener and score reader while they process the musical information. Information processing is not free of energy considerations because it entails the transfer of both work and heat.

A third currency is time, although along this dimension, there is neither a source nor a recipient. Information involves work and heat in its production, copying, and transformation. These activities do not take place instantaneously, but rather over a period of time. The amount of time is variable and depends on the machinery and circumstances. A performance of Beethoven's Ninth Symphony takes about an hour. The software that encodes the performance required several years for the composer to develop. Copying the software to a computer hard drive requires a few seconds at most. Information processing is not free on account of work and heat requirements. That these resources are expended over finite time intervals is also a critical feature.

There is another thread in the net by way of statistics. Information accommodates digital representations. Each element of a message generally appears with a characteristic frequency. The elements are not independent of one another. The vowels a, e, and i, for instance, appear more often in printed text than the consonants w, x, and z. When the letter q appears, it is almost always followed by a u. Music offers a multitude of examples. When Beethoven wrote in the key of C-minor, he committed the note E-flat to paper more often than E-natural. When he included E-flat in a measure, he typically placed G-natural somewhere else in the measure. The characters assembled in digital formats pose natural frequencies and cluster patterns. Together these underpin the structure necessary for transmitting, receiving, and making sense of musical information.

Information is statistical and thus possesses a facet that is probabilistic. A reader of English texts knows and thus anticipates many sentences to begin with T, and that the symbol will be followed by he. A performer of Beethoven's music expects certain notes and sequences dependent on the key signature. In text and music, certain units demonstrate higher occurrence frequencies than others. Yet more than elementary probability weighs in on analysis and discussion. The concepts of conditional and joint probability are also brought to the table.

Examples from everyday life illustrate quite a few properties of information. Chemistry classrooms and labs contribute their share as well. Texts and music scores offer digital reductions of neurological signals and acoustic waves. Molecular structure diagrams serve an analogous purpose for conveying Angstrom-scale packages of electric charges. Cyclohexanone is one of several stable packages with the formula $C_6H_{10}O$. The real ketone in a laboratory flask is an assembly of 54 electrons, 54 protons, and (typically) 44 neutrons. While the assembly details are nontrivial, the information can be represented in digital terms via:

Note the methodology to employ nontunable elements: the letters, lines, and numbers mean the same regardless of size, shading, and font. The precise combination represents on some level the electronic messages carried by the real molecule. The method succeeds because the packaging is subject to the rules of atomic valence and chemical structure theory. Not just any arrangement of one oxygen, six carbon, and ten hydrogen units communicates the electronics of cyclohexanone. The same building blocks pieced in other ways reflect altogether different messages, stable and not so stable, for example:

The *Handbook of Chemistry and Physics* lists 18 molecules portrayed by these same units [14]. The *Aldrich Handbook of Fine Chemicals* lists 26 structural isomers of cyclohexanone [15].

The digital messages in chemistry, as in written text and music, permit more than one representation. Thus cyclohexanone can be depicted as

or

There are other ways with varying emphases and economy of ink. The digital representations are not confined to two dimensions (2D). Ball-and-stick and space-filling assemblies capture the molecular information in a three-dimensional (3D) way. Computer graphics portray the electronics in 2D by transmitting the illusion of 3D.

It is both appreciated and cursed that information admits ready copying and dispersal. Books, journals, and MP3 files drive this point home. So do everyday molecules. A 1.00 milliliter sample of liquid cyclohexanone contains approximately 5.8×10^{21} copies of the same electric charge package, ignoring isotope considerations. Further, the ketone can be prepared along numerous pathways, for instance, beginning with cyclohexanol. It is straightforward for a chemist to assemble enormous numbers of the cyclic ketone, each communicating the same root message. The enterprise of chemistry succeeds because of high-fidelity copies of electronic information prepared in astronomical-size numbers.

Issues of copying overlap with mutability. Cyclohexanone carries electronic information and is obtained by the chemist transforming other molecules, for example:

There are work, heat, and time resources necessary for synthesizing cyclohexanone, regardless of the source materials. There are free energy losses and enthalpy changes for all the aforementioned reactions. There are work, heat, and time costs of information processing by the chemist in the lab. He or she must expend electrical and magnetic work, and dissipate heat when using an NMR spectrometer and gas chromatograph to distinguish cyclohexanone from the starting materials. Work and heat are integral to the chemical procedures (e.g., bromine titration, NMR analysis), which discriminate cyclohexanone from its enol tautomer.

Information mutates on its own if given the chance. If the chemist mixes cyclohexane, cyclohexene, and bromine in a dark room, a reaction occurs involving the latter two compounds. All the while, the cyclohexane molecules play the role of spectators. The reactions transpire simply because they can—free energy is lost by the solution and new entropy is generated. Yet a second critical reason is that there is discrimination shown during each thermal collision. Each package of electric charge bumps against another along myriad trajectories at a rate approaching 10^{12} sec^{-1}. With every contact, there is an electronic interaction that depends on the atom/covalent bond network of the different parties. The cyclohexene molecules possess a reactive functional group; the cyclohexanes do not. The functional group predicates a different set of interactions during collisions, which can be communicated in digital code formats C-C, C=C, C-H, and so forth. The initiation and discrimination of all reaction pathways take place by the processing of Angstrom-scale information.

Information maintains a statistical thread throughout the atomic and molecular realm. The elements H, He, Li, and so forth are distributed on the planet with spatial frequencies dependent on the source, say, coal mined in West Virginia as opposed to water in Lake Michigan. The frequencies are by no means independent of one another. If an H atom appears in a molecule extracted from coal, odds are that C, O, or N units, and not Li, will be a spatially nearest neighbor.

Several basic properties of information are being illustrated in this chapter, including ones that concern molecules. It is important that information links to chemical systems at the macroscopic scale as well. Let a chemist prepare liquid cyclohexanone in equilibrium with its vapor at temperature 294 K. Let the chemist then isolate the vapor portion. The resulting material will be ever changing at the microscopic level given the molecular translations, rotations, vibrations, and thermal collisions. Yet the macroscopic level affords surprisingly compact representations via essential information. With rare exception, the chemist needs only to measure and record three quantities such as pressure (p), volume (V), and temperature (T). These are the notebook entries that would enable a colleague to construct a system with identical macroscopic properties. In effect, the thermodynamic specifications in three areas are required for successful duplication; any more data would be superfluous for large-scale samples. The variables are tunable by the chemist transferring heat or work to the cyclohexanone molecules. Thus, the information affiliated with the system is transformable virtually without limit. Since the measurements of p, V, and T transpire at finite resolution, each quantity can be represented in finite digital terms. Specifying the state variables such as p, V, and T provides software for replicating and transforming the real physical system.

The properties of information apply regardless of sample size and shape. Work and heat exchanges plus time are necessary for producing and processing the macroscopic information. To ascertain p—obtain information about it—the chemist must allow the cyclohexanone vapor to push down on the fluid of a barometer or vice versa. Alternatively, he or she can measure the electrical or thermal conductivity of the gas. Regardless of method, the chemist and apparatus must expend work in order to purchase information about the system. There are heat exchanges between the vapor and measuring devices if the material is not thermally equilibrated beforehand. Even with equilibration, there is friction internal to a barometer and heat dissipation in the fluid motion. Transactions of work and heat are mandatory to obtaining the pressure information; the same holds true for other quantities of state such as the volume, temperature, and n number of moles of cyclohexanone.

Information at the macroscopic level accommodates transformations. If the chemist compresses the vapor in a leak-proof container while holding the temperature constant, such would yield a new system with new information. As per usual, the costs of mutation are not zero in work, heat, and time. Further, digital methods permit more than one representation. The macroscopic state can be described via p, V, and T expressed in the International System of Units (SI units). Other sets such as atmosphere, liter, and Rankine offer equivalent information, no more or less. They enable preparation of exact replicas of the system.

Yet the different representations extend beyond the unit choices. Cyclohexanone vapor at specified p, V, and T is described just as effectively by, for example, the variable sets

{pressure, volume, moles of molecules}
{temperature, volume, moles of molecules}
{pressure, volume, mass}

Every quantity is measurable at finite resolution and thus admitting of digital reduction. Clearly, there is more than one way to represent the system information—sufficient to allow another chemist to construct a replica. There is a catch in that at least one of the variables must be extensive. The extensive variables hinge on the sample size and amount, whereas the intensive ones do not. This is not too imposing a restriction given that p and T are intensive; V, n moles, and m grams are extensive. How many ways are there to choose three from a finite list and have at least one be extensive? The answer is the objective of Exercise 1.7, at the end of the chapter.

Suffice to say that it is straightforward to assemble variable sets that meet the restriction. There are more variables on the palette such as internal energy (U) and entropy (S), both extensive; and density (ρ) and chemical potential (μ), both intensive. But one should note that not all system variables are easy to measure directly. The pressure of a vapor is obtained via a McCleod or other suitable gauge. The entropy of cyclohexanone vapor does not offer such an immediate handle.

It is important that even the variables themselves allow more than one representation. Symbol p stands for pressure and quantifies the force per unit area exerted by the vapor; T performs likewise for temperature and determines the direction of heat flow if the system is placed in contact with another. In an intriguing way, both quantities are equivalent to differential functions:

$$p = -\left(\frac{\partial U}{\partial V}\right)_{S,n} \tag{1.1}$$

$$T = +\left(\frac{\partial U}{\partial S}\right)_{V,n} \tag{1.2}$$

The functions are themselves represented by alphabetic (nontunable) symbols. They are implied to be continuous although their laboratory measurement is limited to finite resolution.

The information expressed by macroscopic states is of a statistical nature as well. Fluctuations are a feature of every state, equilibrium and not so. Thus, when a chemist specifies the state of cyclohexanone vapor via p, T, and V, he or she knows full well such values are not rock solid. When the molecules collide, they adhere momentarily to one another. The molecules adsorb to and desorb from the walls. Processes such as these cause the sample pressure to rise and fall rapidly and interminably. There are shape fluctuations and, as a result, volume changes of the glass container as well. If the walls are diathermal, there are heat exchanges between the vapor and the surrounding environment; adiabatic walls limit the exchanges to the work of shape changes. The message is that any of the p, T, and V values recorded for the system occur with a certain frequency and fluctuate around averages. Moreover, values other than the averages can be anticipated with a certain probability. The vapor state variables are not independent of one another. If p rises or falls, it affects the density ρ. The information of any system is probabilistic in nature. It is also subject to correlations and constraints. There will be much more to say about equilibrium states, fluctuations, and probability in subsequent chapters.

In the discussion so far, several threads of the information net have been mentioned, most of them fairly simple. One should not be lulled into thinking, however, that the threads are uniformly obvious and straightforward. As the first paragraph implied, a discussion of information becomes quickly complicated. One reason is that the value and action of information depend on the receiver. There is information in the score of a Beethoven symphony. And many individuals are capable of appreciating the black dots, lines, and digits on the printed page. Yet the impact depends on whether the viewer is a conductor of an orchestra or of a passenger train. Equivalent statements can be made about the formula diagrams for molecules. The information can be processed if there is already certain information stored in the receiver—the system placed in contact with the source.

The complexities only begin there. Information costs work to produce, copy, and process. Yet, strangely it can function as a type of fuel itself—one capable of producing, copying, and transforming additional information. One gathers this from: mcules crry info. Text such as this grants another system a capacity to write copy, and process the message in complete form. The same holds for formula diagrams. A chemist can make sense of this chapter's portrayals of cyclohexanone. The chemist knows at once from each diagram that the atom constituency is $C_6H_{10}O$.

Information issues are equally complicated at the macroscopic scale. Chemist A can prepare a sample of cyclohexanone vapor, and convey the state to Chemist B using p, V, and T quantities. Such information empowers Chemist B to infer the mass of the gas. This is true, however, only if he or she has access to the equation of state. The multiplying of information is generally possible but requires information.

There is another complicating feature, namely, that information demonstrates what can be thought of as level properties. The letter C appears on a page because somewhere in the printing pipeline the ASC II code 01000011 was transmitted by one computer and received by another. In the same vein, C can appear only if 01000011 (more information than a single character) initiates a subroutine that activates a certain configuration of pixels (still more information than an eight-character word). Clearly, information is underpinned by greater information; it demonstrates layers in the manner of oil paint. The quantity of information invariably increases the lower the level.

As processors of information, chemists are fortunate to operate at comparatively high levels most of the time, and do not have to sweat the lower-level details. Formula diagrams offer a case in point. The examples of this chapter portray Angstrom-scale charge packages, but their information is definitely of the high-level variety. The molecules can be represented at lower levels and with greater processing costs using molecular orbital, density functional, and other electronic structure formats. The level properties of information apply to macroscopic systems as well. Classical thermodynamics offers the chemist a comparatively high-level description of a system. Kinetic theory and statistical mechanics offer lower-level portrayals but at the cost of additional work. Note that the term *lower-level* does not necessarily equate with *better*; the appropriate level depends on the challenge at hand. The preparation of cyclohexanone from cyclohexanol starting material can be communicated using high-level diagrams. A description that includes electronic, vibrational, and rotational wave functions and partition functions may only obfuscate the methodology. At the macroscopic scale, an isothermal compression of a gas may be described at the level of classical thermodynamics. An appeal to kinetic theory and statistical distribution functions may not assist matters. The point is that the chemist is usually able to travel and make progress on the basis of high-level information. Information in simple diagrams enables the construction of exact replicas of molecules. Information in p, T, V and other variable sets enables reproduction of macroscopic systems.

This chapter touched upon threads of a complicated net. Let it conclude by reiterating points that chart the direction of subsequent chapters.

1. Information is physical as Landauer declares. And the most accessible handles are digital, statistical, and structure dependent in nature. These properties are critical to the quantification of information presented in Chapter 2.
2. Information production, copying, processing, and transformation entail transactions of work and heat. Information accordingly has a feature that is thermodynamic. If energy is a system's capacity to perform work or transfer heat, then information represents a capacity for controlling the work and heat transactions. These issues are at center stage in Chapters 3 through 5 in connection with the macroscopic scale.
3. Information is physical—and chemical. Molecules carry, transfer, and transform electronic information. The information processing effected by collisions in thermal environments makes chemistry possible. These subjects propel Chapters 6 and 7.
4. Information casts a net approaching 3 billion Google hits. Not every aspect can or should be addressed in one book. Issues regarding time costs, transfer fidelity, and parallel processing receive glancing remarks in Chapter 8.

1.2   SOURCES AND FURTHER READING

The net of information has logged multitudinous books of much thought-provoking variety. Several have been cited already. The author further recommends Lowenstein's *The Touchstone of Life: Molecular Information, Cell Communication, and the Foundations of Life* [16] and Berlinski's *The Advent of the Algorithm: The Idea That Rules the World* [17]. For insights into chemical information networks of early days, by all means read *The Invention of Air* by Johnson [18].

1.3   SUGGESTED EXERCISES

1.1 The chapter opened with the statement "Information motivates much discussion." Several declarations followed. Choose one and write a two- to three-page response paper. The response should argue the merits and deficiencies regarding information.

1.2 This chapter presented the idea that information represents a system's capacity for controlling work and heat transactions. As in the first exercise, compose a response paper that addresses the merits and deficiencies of the idea.

1.3 This chapter cited the combining of letters *a*, *e*, and *t* to form messages. Some of the messages are closely related in meaning such as *eat* and *ate*, whereas others are different, for example, *tea*. In a parallel way, atoms combine to form molecules, each carrying an electronic message.
    For this exercise, identify the stable messages that can be assembled using five carbon, eight hydrogen, and two oxygen atoms. These should include the structural and valence isomers of $C_5H_8O_2$ as well as van der Waals dimers. Which combinations are closely related by chemical functionality?

1.4 Each of the letters *a*, *b*, *c*, ... *z* composes 1/26 of the English alphabet. Each can function as the first character of a word: ate, bobcat, chemistry, and so forth. (a) Refer to an English language dictionary and identify the fraction of pages associated with each letter as a first character. (b) Arrange the fractions in ascending order, for example, 0.0565, 0.0684, 0.135, and so on. (c) Form a summation based on the ascending fractions: 0.0565, (0.0565 + 0.0684), (0.0565 + 0.0684 + 0.135), and so on. (d) Pair the summed terms, respectively, with 1/26, 2/26, 3/26, and so on. (e) Prepare to construct a plot: the fraction sums on the abscissa and 1/26, 2/26, and so forth on the ordinate. (f) Write in advance the character (linear, exponential, etc.) anticipated for the plot. (g) Construct the plot by hand. Does the plot match expectations? Please discuss.

1.5 The *Handbook of Chemistry and Physics* functions as an abridged molecular dictionary [14]. It includes a tabulation of entries for a given formula, for example, 10 entries correspond to $C_4H_8Cl_2$. (a) Count the number of entries associated with a single carbon atom. Do likewise for entries for two-carbon, three-carbon, and so forth. (b) Prepare to plot the number of entries versus carbon atoms on the ordinate and abscissa, respectively. (c) Write in advance what the plot is anticipated to look like. (d) Construct the plot. Does it match expectations? Please discuss.

1.6 Describe two examples drawn from chemistry where probability plays a role. Do likewise regarding conditional probability.

1.7 The macroscopic state of a one-component system can be specified via three variables as long as at least one is extensive. Consider a palette of the following: p, V, T, n moles, mass m, density ρ, internal energy U, molar volume $\bar{V}$, and entropy S. How many valid combinations can be assembled?

REFERENCES

[1] Misner, C. W., Thorne, K. S., Zurek, W. H. 2009. John Wheeler, Relativity and Quantum Information, *Physics Today*, April, p. 44.

[2] Landauer, R. 1987. Computation: A Fundamental Physical View, *Phys. Scr.* 35, 88; Landauer, R. 1993. Information Is Physical, *Proc. Workshop on Physics and Computation*, *IEEE Comp. Sci. Press*, Los Alamitos, p. 1.

[3] Hawking, S. W. 1975. Particle Creation by Black Holes, *Commun. Math Phys.* 43, 199; Hawking, S. W. 2005. arXiv:hep-th/0507171v1, Information Loss in Black Holes.

[4] Susskind, L. 2008. *The Black Hole War: My Battle with Stephen Hawking to Make the World Safe for Quantum Mechanics*, Little Brown, New York.

[5] Marvin, C. 1987. Information and History, in *The Ideology of the Information Age*, Slack, J. D., Fejes, F., eds., Ablex Publishing, Norwood, NJ.

[6] Küppers, B. O. 1990. *Information and the Origin of Life*, Section II, MIT Press, Cambridge, MA.

[7] Tribus, M., McIrvine, E. C. 1971. Energy and Information, *Sci. Amer.* 225, 179.

[8] *Webster's Third New International Dictionary of the English Language Unabridged.* 2002. Gove, P. B., ed. in chief, Merriam-Webster, Inc., Springfield, MA.

[9] Hayles, N. K. 1999. *How We Became Posthuman*, chap. 3, University of Chicago Press, Chicago.

[10] Wiener, N. 1961. *Cybernetics*, p. 10, MIT Press, Cambridge, MA.

[11] The Rolling Stones. 1972. *Hot Rocks 1964–1971*, side one, track four.

[12] Bilton, N. 2009, December 10. Part of the Daily American Diet, 34 Gigabytes of Data, Business and Technology Section, *The New York Times*. The article reports research carried out by the Information Center at the University of California, San Diego.

[13] Cornish-Bowden, A. 2004. *The Pursuit of Perfection: Aspects of Biochemical Evolution*, chap. 6, Oxford University Press, Oxford.

[14] Weast, R. C., ed. 1972. *Handbook of Chemistry and Physics.* Chemical Rubber Co., Cleveland, OH.

[15] *Aldrich Handbook of Fine Chemicals.* 2006. Sigma-Aldrich Company, St. Louis, Mo.

[16] Loewenstein, W. R. 1999. *The Touchstone of Life: Molecular Information, Cell Communication, and the Foundations of Life*, Oxford University Press, New York.

[17] Berlinski, D. 2000. *The Advent of the Algorithm: The Idea that Rules the World*, Harcourt, New York.

[18] Johnson, S. 2008. *The Invention of Air*, Riverhead Books (Penguin Group USA), New York.

2 A Quantitative Look at Information

Information is examined in quantitative terms. The ingredients necessary for quantifying are described, as are the units and interface with probability. The examples include ones based on aromatic substitution and peptide chemistry. Three types of information are illustrated. Peptide chemistry and mass spectrometry are used to establish three probability functions of wide applicability. Several tools of probability are described to close the chapter. Suggested exercises follow.

2.1 ESSENTIAL INGREDIENTS

The first chapter looked at information in the qualitative sense. A few of the exercises ventured a short distance into quantitative territory where Chapter 2 plants roots. We begin with the ingredients essential to realizing information in a quantitative way. The first of these is a venue that expresses a finite number of well-defined states. There is a lot of room to operate here, as *state* can be defined loosely as a "way or condition of being." The face of a coin describes a way for the coin to reside on a flat surface. The face marks one state of the coin. Likewise, the diagram

describes one of the stable arrangements—and thus a viable state—of a molecule with formula $C_9H_{11}NO_2$. Even the word *stable* represents a state because it is one method for six letters to be arranged. The concept of state applies to venues everyday and rarefied. One can argue that it applies to all venues, real and imaginable.

The second ingredient is a mechanism for the state to communicate an unambiguous message. The viewer knows the coin's face by the way it reflects visible light. The optical pattern of the heads (H) state differs from that of tails (T). The previous assembly of nine carbon, eleven hydrogen, one nitrogen, and two oxygen atoms differs from structural isomers such as

Obviously, the word *stable* differs from *tables* in light-reflecting properties and English language message—the two words mean very different things. Note that the state and message, whether derived from coins, molecules, or printed text, must be digital in nature. Thus, the essence of the state must be immune to changes in size, shading, and other incidentals. This is clear in the examples. The H message of a 25-cent coin is invariant to changes in the age, mint source, or metallic content. All of the following refer to 3-nitro-4-ethyl-toluene irrespective of orientation and abbreviation details:

The message in *stable* is likewise not qualified by the style, color, or paper used for representation. In short, for information to be quantifiable, a certain robustness is necessary for the states and messages.

The third ingredient is also mechanistic. There must be some uncertainty to the state message before the receiver accesses it. A coin at rest offers two ways of being with messages communicated by visible light reflection. There is the time-honored method at the start of football games, to access the message with uncertainty as to the outcome. The referee has only to toss and rotate the coin, allowing it to land freely.

If the previous nitroaromatic was generated by collisions between NO_2^+ and 4-ethyl-toluene

there would be two routes for the chemistry. The possible reaction products are communicated by the following diagrams:

In the left-side case, the nitro group is *ortho* to the methyl group; in the right it appears *meta*. This difference allows for abbreviation of the states and messages, namely, *ortho* and *meta*, with the methyl group specified and understood as the point of reference. Note that there would be abundant sources of uncertainty regarding the reaction product. During the chemistry, the molecules rotate like coins, only much faster and with more complicated trajectories. Regarding letter arrangements, it is straightforward to program a computer to print the letter *s* followed by *table* or vice versa. There is uncertainty in the outcome if a random number generator governs the state choice. For instance, the program can examine the fractional portion f of $(\pi + u)^5$. Here u serves as a seed quantity with value <1, and π is the transcendental number 3.1415926535... . For example, with $u = 0.24589395105$, $(\pi + u)^5 = 446.0545076918...$ and $f = 0.0545076918$ after truncation. The computer can be programmed to print *stable* for $f \le 0.50000000$ and *tables* otherwise. Note the dichotomy. The outcome is dictated not by random events per se but rather the mathematical details of $(\pi + u)^5$. But an outcome is unpredictable from the user's standpoint given arbitrary choices of u and the number of decimal places carried. The computer provides what is essentially a coin flip by alternate means. Note that the choice of the exponent in the random number generator is arbitrary and need not even be an integer. The production of random numbers merits and accommodates multiple strategies.

In summary, information in the quantitative sense requires a venue that poses a finite number of states, each able to convey an unambiguous message. There must be a mechanism for accessing the message with uncertainty beforehand. Let the chemist flip a quarter and inspect the result. Let he or she arrange for 4-ethyl-toluene and the electrophile NO_2^+ to interact in solution, and then to assay the product via spectroscopy. Let the chemist seed and operate the random number generator and printer. Let the chemist inspect the final results, which turn out to be:

$$\text{T} \qquad meta \qquad tables$$

He or she has thereby trapped—irreversibly acquired—information. But then the question is how much.

Three different venues and states are being discussed—coins, molecules, and letter arrangements. It is important that the states are all amenable to the same brand of labeling using the digits 0 and 1 as spelled out in Table 2.1. The choice of which digit pairs with which state is arbitrary. So too is the use of 0 and 1 as labels. *A* and *B*, α and β, *red* and *black*, and so forth work just as well. The point is that the label for a state fits and can reside indefinitely in a single slot. Flipping a coin is tantamount

---

TABLE 2.1

Venues, Allowed States, and Digital Labels

Venue	Allowed States	Labels
Coin Flipping	H, T	0, 1
Reagent Collisions	*ortho, meta*	0, 1
Letter Arrangements	stable, tables	0, 1

---

to converting a slot initially devoid of a label to one that is permanently filled, for example, [ ] → [H]. Equivalent statements apply to all information venues irrespective of the label choices.

There are several points that follow, the first being that the absence of information equates with uncertainty about which label will end up filling a slot. Experiments offering results that are foregone conclusions may be attractive for one reason or another, but they offer zero information in the quantitative sense. The second point is that all—not merely some—of the possible states and labels must be known in advance. Venues whereby only some of the states are acknowledged and in play demand further groundwork for information to be quantified. Third, the amount of information equals the number of slots needed to label the messages efficiently, that is, with a minimum of leftovers. For each of the Table 2.1 venues, a single slot is required; the process of registering a message results in a single binary digit of information. In more compact terms, each venue offers 1 bit of information, the word *bit* serving as an abbreviation for "binary digit."

The fourth point is that venues and states can be combined. If the chemist flipped a quarter on the lab bench while allowing 4-ethyl-toluene and NO_2^+ to interact, the combination exercise would offer 2 bits of information. The joining of all three venues of Table 2.1 yields 3 bits. Information is an additive quantity.

The fifth point is a nuanced one. If the chemist were to combine all three venues, a possible state and binary label outcome would be:

H	*ortho*	*tables*
0	0	1

while another would be:

T	*ortho*	*stable*
1	0	0

It is straightforward to verify eight combination-states with the label sets as follows:

Coin Flip	Aromatic Substitution	Letter Arrangement
0	0	0
0	0	1
0	1	0
0	1	1
1	0	0
1	0	1
1	1	0
1	1	1

Therein lies the point. Information in the quantitative sense does not refer to the state message itself but rather to its length. Experiments that result in the 011 outcome—or 010, 110, and so forth—yield the same amount of information, namely, 3 bits. It is easy and even tempting to overlook this property. Information in a facts and data context is synonymous with message. People gather information from newspapers, books, Web sites, and so on; pets receive information from cues of their owners. In matters quantitative, however, the concept connects with the size of a message. To go one step further, one observes that for the Table 2.1 venues, whether single or combined, the number of possible state messages Ω is an integer power of 2:

$$\Omega = 2^I \tag{2.1}$$

The exponent I equates with the binary digit slots required to label the state message. Therefore,

$$\log_2 \Omega = I \tag{2.2}$$

which is the information measured in binary digits. Note that bits is not the only unit that can attach, but it is the most frequent and popular. This is in spite of real-life venues rarely posing Ω equal to an integer power of 2. But there arises a minor quandary. Handheld calculators are lacking in log-base-2 buttons. How does the chemist compute information in more general scenarios?

The answer is that information I always hinges on the number of possible states Ω. If one considers:

$$\Omega = e^y = 2^I \tag{2.3}$$

it follows that,

$$\log_e \Omega = y = \log_e(2^I) = I \cdot \log_e(2) \tag{2.4}$$

In turn, I is quantified in bits—as opposed to other units—by the relations:

$$I = \frac{\log_e \Omega}{\log_e (2)} \approx \frac{\log_e \Omega}{0.693} \tag{2.5}$$

Thus for $\Omega = 5$, 89, and 1012, I equates with approximately 2.32, 6.48, and 9.98 bits, respectively. The major point is that information amounts always arrive via logarithm functions. In the base-2 system, the relevant unit is bit. In base-e units, the unit nit attaches while nat is an alternative. A perusal of any dictionary will show why nit is not a pleasant-sounding term. Yet regardless of unit, the additivity of information is consistent with the logarithm function. If the chemist conducts three independent experiments, each with possible states Ω_1, Ω_2, and Ω_3, then the total combined number of possible states (Ω_{total}) is:

$$\Omega_{total} = \Omega_1 \cdot \Omega_2 \cdot \Omega_3 \tag{2.6}$$

The sum total information I_{total} is then,

$$\begin{aligned} I_{total} &= \log_2 (\Omega_1 \cdot \Omega_2 \cdot \Omega_3) \\ &= \log_2 \Omega_1 + \log_2 \Omega_2 + \log_2 \Omega_3 \\ &= I_1 + I_2 + I_3 \end{aligned} \tag{2.7}$$

The total number of states is obtained via multiplication, whereas the total information is the result of addition. Logarithm functions support state and information properties in an elegant way.

2.2   THE INTERFACE OF INFORMATION WITH STATE LIKELIHOOD

The first section approached each state of a venue as equally likely; one message was anticipated as much as another. This seemed reasonable for coin flipping and using random number generators to determine the printing of *stable* versus *tables*. The assumption of equal likelihood is not so justified in reactions of 4-ethyl-toluene. After all, the two substituents pose different steric effects and activate the aromatic ring electronically to different degrees. The frequency of nitration *ortho* to the methyl group is anticipated to be different from the *meta* alternative. How much different depends on factors such as temperature and solvent. The upshot is that nature's preference for one outcome over another means that there should be less uncertainty about the reaction product. The reduced uncertainty about two possible states means less information for the chemist—something less than 1 bit. But then, how much less?

One intuits that if the state likelihoods were only slightly skewed, say, the *ortho* product was 1.10 times more likely than *meta*, there would be only slightly less information than 1 bit available. If instead, the *ortho/meta* likelihoods were dramatically skewed, say, by a factor of 10, the available information would be considerably less than 1 bit. If, for any reason, the imbalance were a factor of 10^4, the information would be reckoned as slightly above zero bits.

An imbalance factor F ($= 1.10, 10, 10^4$) is readily converted to a pair of fractions, f_{ortho} and f_{meta}:

$$f_{ortho} = \frac{F}{F+1} \tag{2.8}$$

$$f_{meta} = 1 - f_{ortho} \tag{2.9}$$

The results then approximate the fraction of times—the likelihood—the particular state will be realized in experiments. For the aforementioned three cases, one has

$$F = 1.10 : f_{ortho} \approx 0.5455, f_{meta} \approx 0.4545 \tag{2.10}$$

$$F = 10 : f_{ortho} \approx 0.9091, f_{meta} \approx 0.0909 \tag{2.11}$$

$$F = 10^4 : f_{ortho} \approx 0.9999, f_{meta} \approx 0.0001 \tag{2.12}$$

A more formal connection between the fractional likelihoods and probabilities is reserved for Section 2.4. For now, it is sufficient to note that formulae analogous to Equations (2.8) and (2.9) extend beyond aromatic substitution. The generality is conveyed by labeling the likelihood fractions by subscripts 1 and 2: f_1 and f_2. Then, if one considers the fractions as logarithm arguments and multiplies the results by -1, one arrives at:

$$-\log_2 f_1 = \frac{-\log_e f_1}{\log_e(2)} \approx \frac{-\log_e f_1}{0.693} \tag{2.13}$$

$$-\log_2 f_2 = \frac{-\log_e f_2}{\log_e(2)} \approx \frac{-\log_e f_2}{0.693} \tag{2.14}$$

This leads to Table 2.2. Included is the case of the equal-likelihood venues of Section 2.1 where $f_1 = f_2 = 0.5000$. The purpose of Table 2.2 is to demonstrate that the two rightmost columns move in opposite directions. The term $\frac{-\log_e(f_i)}{0.693}$ increases as f_i inches toward zero; the term approaches zero as f_i moves closer to 1.

One meets another important quantity along intuitive lines. Whereas information scales with the uncertainty reduced by an experiment, the term $\frac{-\log_e(f_i)}{0.693}$ is tied to the degree of surprise of an observation. Unlikely events, such as winning Illinois powerball lotteries or nitration of α,α,α-trifluorotoluene to yield an *ortho* product, have fractional likelihoods that border on zero. Such events predicate large surprises in the very literal sense. Failing to win a lottery or observing nitration *meta* to the CF_3 site have fractional likelihoods very close to one [1]. These outcomes generate

TABLE 2.2

Fractional Likelihoods and Surprisals

f_1	f_2	$S_1 = \dfrac{-\log_e(f_1)}{0.693}$	$S_2 = \dfrac{-\log_e(f_2)}{0.693}$
0.5000	0.5000	1.000	1.000
0.5455	0.4545	0.8745	1.137
0.9091	0.0909	0.1375	3.459
0.9999	0.0001	0.01442	13.29

virtually no surprise at all. One gathers that the surprisal quantity (S_i) allied with state i having fractional occurrence (f_i) is established by:

$$S_i = -\log_2 f_i \approx \frac{-\log_e f_i}{0.693} \tag{2.15}$$

Now *every* possible state contributes to the information of an experiment. In fact, the contribution is weighted by the state likelihood imbedded in the surprisal. One intuits that,

$$-f_i \log_2 f_i \approx \frac{-f_i \log_e f_i}{0.693} \tag{2.16}$$

is the contribution to the information by the ith state. It is a short step away to find that the total information I, given N possible states, follows from a sum of weighted surprisals, namely,

$$I = -\sum_i^N f_i \log_2 f_i$$

$$\approx \frac{-1}{0.693} \times \sum_i^N f_i \log_e f_i \tag{2.17}$$

By applying Equation (2.17) to the Table 2.2 scenarios, one obtains Table 2.3. The latter shows how the fractional likelihoods together impact the information of an experiment. A fractional likelihood by itself bears on the chemist's uncertainty—or lack of it—of observing a particular state: if $f_i = 0.980$, the chemist is fairly certain that the ith state will rear its head in the experiment; if $f_i = 0.485$, the chemist is quite uncertain about the state's next manifestation. Information, by contrast, is founded on the collective uncertainty that involves all the states 1, 2, i, ... , $N-1$, N, likely and not so likely.

Equation (2.17) and Table 2.3 make the case that the information available from an experiment is a weighted average of the surprisal terms. In statistical terminology,

---

TABLE 2.3

Fractional Likelihoods, Surprisals, and Summations

f_1	f_2	$f_1 \cdot S_1$	$f_2 \cdot S_2$	$\dfrac{\sum\limits_{i=1}^{N=2} f_i \cdot S_i}{0.693}$
0.5000	0.5000	0.5000	0.5000	1.000
0.5455	0.4545	0.4770	0.5170	0.9940
0.9091	0.0909	0.1250	0.3145	0.4395
0.9999	0.0001	1.442×10^{-4}	1.328×10^{-3}	1.472×10^{-3}

---

information is the *expectation* of the surprisal. Note as well from Table 2.3 that the information is greatest when the state likelihoods are equal. This is a truism that extends beyond venues featuring only two states; the conditions that yield the most information are ones where all the states manifest equal likelihood. Figure 2.1 illustrates this explicitly for binary venues where $f_1 + f_2 = 1$; $f_2 = 1 - f_1$. Clearly the maximum information, namely, 1.00 bit, is evidenced when $f_1 = 0.500$. Just as important, the information converges to zero as f_1 approaches 1 or 0. This statement echoes a point made in Section 2.1: Venues posing only one possible outcome are not experimental in the strict sense of the word because they offer zero information. By contrast, the maximum information follows from situations where there is maximum uncertainty about the outcome.

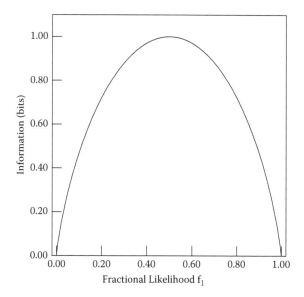

FIGURE 2.1   Information for a two-state venue as a function of fractional likelihood f_1. The information is maximum when states 1 and 2 occur with equal likelihood.

Table 2.3 illustrates another property albeit implicitly. It is that experimental venues that pose information are affiliated with complete sets of fractional likelihoods. A set can host a few or many members. The number is always dictated by the possible states. The following are examples of viable and thus complete sets { f_i }: {0.5000, 0.5000}, {0.9500, 0.0500}, {0.9500, 0.0300, 0.0200}. It is sometimes the case that two or more f_i are equal, for example, {0.8800, 0.0300, 0.0300, 0.0600}. What is always the case is that the f_i sum to 1, as introduced by f_{ortho} and f_{meta} in Equations (2.8) and (2.9). Information venues are grounded upon these sets—collections of fractions that sum to 1. In turn, quantifying information begins with establishing a complete set of fractions. Once the set is known, the information in bits, nits, or other choice of units follows straightaway.

The preceding idea shows the applicability of information in chemistry, if not scientific fields in general. If an experiment ascertains whether a molecule is a D- or L-isomer, there is 1 bit of information acquired if each isomer is equally likely in occurrence. If an assay establishes the first base unit of a DNA sample, 2 bits of information are typically obtained. If a thermometer registers whether the temperature of a sample is below 273 K, 1 bit is trapped. Information in the quantitative sense follows from interfacing an experiment with queries that admit *yes or no* answers. The childhood game of twenty questions directs the inquiring party toward a conclusion based on a maximum of 20 bits of information. Information is the quantity in search of systems, inquiries, and experiments. One notes the resemblance of Equation (2.17) to the entropy of mixing for an ideal solution [2]:

$$\Delta S_{mix} = -n_1 \cdot R \cdot \log_e(X_1) - n_2 \cdot R \cdot \log_e(X_2) - \cdots$$

$$= -R \cdot \sum_{j=1}^{\kappa} n_j \cdot \log_e(X_j) \tag{2.18}$$

where R is the gas constant, and n_j and X_j are the respective mole amounts and fractions for each component of a system hosting κ total. Just as important to note is the entropy of a system established by statistical mechanics [3]:

$$S = -k_B \cdot \sum_j prob(j) \cdot \log_e(prob(j)) \tag{2.19}$$

Here k_B is Boltzmann's constant and $prob(j)$ is the probability of observing the jth state of a system. The similarities of Equations (2.17) through (2.19) are not coincidental. It is apparent that information and entropy are related if not alternate sides of the same coin. The inaugural properties and applications of Equation (2.17) were the brainchild of Claude Shannon and thus I is commonly referred to as the *Shannon information* [4]. The term *Shannon entropy* is written almost as often on account of the ties to Equations (2.18) and (2.19). The mixing entropy of Equation (2.18) is visited several times in subsequent chapters.

2.3   THE ROLE OF PROBABILITY

Real-life experiments—tossing coins, aromatic substitution, and so on—pose states and fractional likelihoods (f_i). The latter are usually established by weighing the number of times (N_i) the ith state is registered against the total number of observations (N_{total}):

$$f_i = \frac{N_i}{N_{total}} \tag{2.20}$$

Yet f_i encountered in Tuesday's experiments will generally not match Wednesday's. Section 2.2 offered that f_i are typically different for different states. In the same vein, a given f_i is subject to fluctuations. How do these issues square with information?

The short answer is that information, as illuminated by Shannon and others, is formally based on probability. The latter is an idealized extension of f_i; the probability $prob(i)$ associated with the ith state equates with f_i in the limit of infinite trials or observations, that is,

$$prob(i) = \lim_{N_{total} \to \infty} \left(\frac{N_i}{N_{total}} \right) \tag{2.21}$$

Infinite observations are impossible in the chemist's lifetime; this is a first idealization of $prob(i)$. Querying states independently one at a time is also not always feasible; this is a second source of idealization. Probability ideas reach far nonetheless. One looks to $\psi_{2s}^2 d\tau$ for an example. This is interpreted in chemistry classes as the probability of observing an atom's $2s$ electron in an infinitesimal volume element ($d\tau$), as dictated by a wave function (ψ_{2s}). Such a probability is not very accessible to the chemist, experimentally at least. In spite of the observation complexities, however, probability concepts are applied widely. At the minimum, they point to critical questions for the chemist to consider.

A rigorous discussion of probability begins with set theory. On simpler ground, the tools of probability can be acquired by thinking exercises. The time-honored ones appeal to balls drawn from urns, poker hands, thrown darts, and tossed coins [5]. Yet the exercises need not be so macroscopic in character. Spin populations and electron clouds have also been used to illustrate and thus reinforce probability concepts [6,7]. The microscopic nature of such examples makes them easier to imagine than to access by experiment.

This section will add to the list by considering peptides—molecules formed by the covalent linking of amino acids. Their applicability derives in several respects. A peptide's primary structure is conferred by the amino acid sequence. The allowed states are countable based (typically) on 20 building blocks. A peptide's states are easy to illustrate and label. Table 2.4 presents the names and abbreviations of the standard amino acids along with formula weight data. Nature employs quite a few more, but the standard 20 suffice for most purposes. Figure 2.2 illustrates a few amino acids and a tripeptide in formula diagram terms; issues of stereochemistry are completely ignored. Peptides are obtained from classical synthesis, robotic, and

TABLE 2.4
Naturally Occurring Amino Acids

Amino Acid	Abbreviations	Formula Mass (grams/mole)
Alanine	Ala, A	89.09
Arginine	Arg, R	174.2
Asparagine	Asn, N	132.12
Aspartic acid	Asp, D	133.1
Cysteine	Cys, C	121.16
Glutamine	Gln, Q	146.14
Glutamic acid	Glu, E	147.13
Glycine	Gly, G	75.07
Histidine	His, H	155.15
Isoleucine	Ile, I	131.17
Leucine	Leu, L	131.17
Lysine	Lys, K	146.19
Methionine	Met, M	149.21
Phenylalanine	Phe, F	165.19
Proline	Pro, P	115.13
Serine	Ser, S	105.09
Threonine	Thr, T	119.12
Tryptophan	Trp, W	204.23
Tyrosine	Tyr, Y	181.19
Valine	Val, V	117.15

cellular technologies [8]. The states are established using Edman-type sequencing, chromatographic methods, and mass spectrometry [9]. The physical properties of peptides are pursued along multiple lines such as molecular weight. There is more to say on this subject in Section 2.4.

For now, we consider a set of tripeptides, one member of which was illustrated at the bottom of Figure 2.2. The first of the thinking exercises calls for cellular machinery that can generate the molecules in various amounts. Let the machinery confine the reagent palette to glycine (G) and valine (V). The possible (and distinguishable) states are GGG, GGV, GVG, GVV, VGG, VGV, VVG, and VVV where the left-to-right sequences indicate the N- to C-terminal direction. Let the production of molecules be in such high numbers that the mole fractions perform double-duty as probability values. Naturally, the state populations will depend on transcription and translation enzymes, G and V availability in the cell, and genetic programming. These details are of no concern here.

Let the set of probabilities be those constructed arbitrarily and listed in Table 2.5. The sum of the weighted surprisals (rightmost column) is 2.485. Then 2.485 bits is the amount of information availed when the chemist randomly selects a tripeptide from the cell and determines the primary structure. If the eight possibilities were produced by the cell in equal numbers, then the information would be 3.000 bits. The lesson of Table 2.5 is presented pictorially in Figure 2.3. The open squares mark the weighted surprisals for the eight states. The plot emphasizes that however

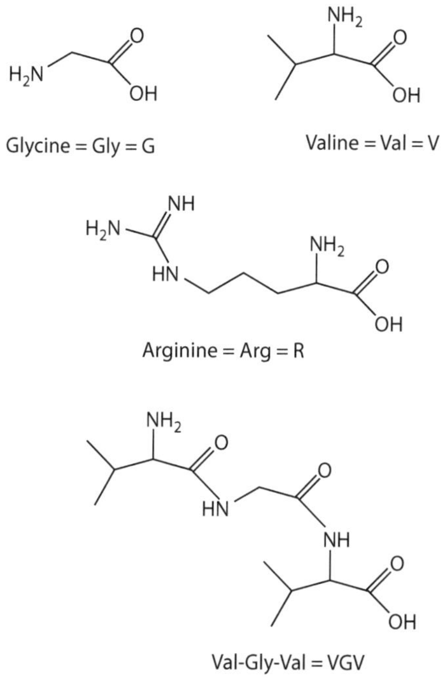

Glycine = Gly = G

Valine = Val = V

Arginine = Arg = R

Val-Gly-Val = VGV

FIGURE 2.2 Formula diagrams and abbreviations of sample amino acids and peptides. Each amino acid and peptide offers three- and single-letter abbreviations.

scattered the surprisal values, the terms are all positive. In turn, their summation only enhances and never detracts from the total information. This is gathered from the heights increasing left to right for the filled squares. Note that this is the case regardless of how the states VVV, VVG, and so forth are numbered and referred to by the chemist. There is more than one way to apply the indices 1, 2, ..., 8.

The states also admit more than one type of investigation. Thus, one relevant set of probabilities can beget others. Each offers its own information amount in careful experiments. For example, a chemist could inquire about the probability of observing

TABLE 2.5

Tripeptide States, Probabilities, and Surprisals

Index i	State	prob (i)	S_i	prob (i)·S_i
1	GGG	0.3000	1.737	0.5211
2	GGV	0.0500	4.322	0.2161
3	GVG	0.0800	3.644	0.2915
4	GVV	0.2000	2.322	0.4644
5	VGG	0.0100	6.644	0.0664
6	VGV	0.2500	2.000	0.5000
7	VVG	0.0200	5.644	0.1129
8	VVV	0.0900	3.474	0.3127

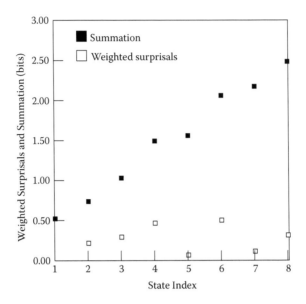

FIGURE 2.3   Weighted surprisals and summation. The plot is based on data of Table 2.5.

a peptide generated by the cell having at least one V: $prob(\geq 1\ V)$. Referring to Table 2.5, such curiosity is followed up by summation and division:

$$prob(\geq 1\ V) = \frac{sum\ of\ probabilities\ of\ states\ that\ have \geq 1\ V}{sum\ of\ probabilities\ of\ all\ states}$$

$$= \frac{0.0500 + 0.0800 + 0.2000 + 0.0100 + 0.2500 + 0.0200 + 0.0900}{0.3000 + 0.0500 + 0.0800 + 0.2000 + 0.0100 + 0.2500 + 0.0200 + 0.0900}$$

$$= 0.7000 \qquad (2.22)$$

Then the probability of observing a peptide lacking in V is simply $1 - 0.7000 = 0.3000$. Information-wise, an experiment aimed at the question "Does the randomly selected peptide contain at least 1 V?" offers Shannon information:

$$I = \frac{-1}{\log_e(2)}[0.700 \cdot \log_e(0.7000) + 0.300 \cdot \log_e(0.3000)] \approx 0.881\ \text{bits} \quad (2.23)$$

There are numerous questions in the same vein. What is the probability that the peptide contains exactly one V? Pursuit of the answer is guided by a fraction set {0.1400, 0.8600} and Shannon information 0.584 bits based on the question "Does the peptide contain exactly one V?" In these simple examples, the probability set of Table 2.5 is used to generate others. Most important, they show the concept of *state* to be a fluid one, and that there can be numerous descendents of a probability set.

Probabilities can have strings attached. For instance, what is the probability that, given that a randomly isolated tripeptide contains V, there are also two Gs? To address this, the chemist must examine the fraction of times that V occurs. The chemist then compares that fraction to the fraction of times two Gs are included. One computes the ratio

$$\frac{sum\ of\ probabilities\ of\ 2\ Gs\ in\ peptides\ containing\ V}{sum\ of\ probabilities\ of\ all\ states\ containing\ V}$$

$$= \frac{0.0500 + 0.0800 + 0.0100}{0.0500 + 0.0800 + 0.2000 + 0.0100 + 0.2500 + 0.0200 + 0.0900} = 0.2000 \quad (2.24)$$

Information-wise, experiments motivated along this dimension offer

$$I = \frac{-1}{\log_e(2)}\left[0.2000 \cdot \log_e(0.2000) + 0.8000 \cdot \log_e(0.8000)\right] \approx 0.722 \text{ bits} \quad (2.25)$$

The idea is that the chemist has somehow already determined that the peptide contains V. His or her follow-up experiment then targets the question "Are there also two Gs present?" A similar question would be "What is the probability that, given there are two Vs anywhere in the peptide, there is also G present?" This probability is:

$$\frac{0.2000 + 0.2500 + 0.0200}{0.2000 + 0.2500 + 0.0200 + 0.0900} = 0.8100 \quad (2.26)$$

The amount of information allied with the corresponding yes–no question and the probability set {0.8100, 0.1900} follows immediately. The features to notice in the last two examples are that they portray the workings of *conditional* probability.

A third type of query arrives by considering combination states. For peptides, such states arise not only from the amino acid composition but also from the sequence, usually interpreted left to right. For example, the molecules in Table 2.5 can each be viewed as expressing two combination states, one marked by the N-terminal (leftmost) unit and the other by the combination of the two rightmost units. The resulting probabilities must then be distinguished via two indices, i and j, and *joint* probability terms $prob(i, j)$. The i index runs only from 1 to 2, referring to G or V, respectively, as the N-terminal unit. The j index runs from 1 to 4 given the respective GG, GV, VG, and VV possibilities. Table 2.5 reports that GGG occurs with probability 0.3000. In the joint probability view, this fraction would equate with $prob(i = 1, j = 1)$. It is straightforward to compute other $prob(i, j)$: $prob(i = 1, j = 2) = 0.0500$, $prob(i = 2, j = 3) = 0.0200$, and so forth. It is interesting to compare joint probabilities assembled via Table 2.5 to ones that would apply in the absence of bias exercised by the cell. If V appeared as often as G for the N-terminal unit, then $prob(i = 1) = 1/2 = 0.5000$. If GG, GV, VG, and VV were all equally likely as the rightmost units then $prob(j = 1, 2, 3, 4) = 1/4 = 0.2500$. In the bias-free case, $prob\,(i, j) = prob(i) \times prob(j) = (1/2) \times (1/4) = 1/8 = 0.1250$. Comparisons of joint probability to values in bias-free scenarios offer the first clues about correlations. In looking at Table 2.5 data, one sees that the presence of V in the middle slot of the tripeptide enhances the chance of V occupying the C-terminal position.

2.4   INFORMATION AND EXPECTATION

The chemist does not labor without expectations. An experiment in which the state probabilities manifest as anticipated is valuable for the currency granted to a particular theory. Results that demonstrate otherwise are still of service and perhaps even more so. They motivate the search for a new theory or modification of the old one.

Information offers a method of weighing expected versus contrary results. The method assigns a cost or penalty in bits for results that arrive in an unanticipated fashion. There is zero penalty charged for outcomes that conform to expectations. The penalties are substantial when the results fail wholesale to match expectations. Matters work as follows.

Suppose that the chemist, by virtue of training and expertise, anticipated that a certain state i would rear its head 40% of the time, that is, $prob(i) = 0.4000$. If indeed such an occurrence frequency were realized experimentally, there would be no penalty attached to the correct-in-advance expectations. The chemist foresaw the state contributing $-0.4000 \log_e(0.4000)/\log_e(2) \approx 0.529$ bits to the weighted surprisal sum, and that was indeed the case. As discussed in Section 2.1, it requires digital code—0/1, A/B, red/black—to label states. Evidently the chemist was well prepared with the correct number of code units purchased and ready to go.

If instead, the fractional occurrence (interpreted as probability) turned out to be 0.100, such a state would contribute only $-0.1000 \log_e(0.1000)/\log_e(2) \approx 0.332$ bits to the weighted surprisal sum. Under these circumstances, the chemist would have overestimated the bits needed for labeling the states. The chemist would have overpaid.

One considers the reverse case. If the chemist had anticipated the fractional occurrence as 0.1000 with weighted surprisal of approximately 0.332 bits, and instead had observed 0.4000 and weighted surprisal 0.529 bits, a deficiency of code would have been encountered. The chemist would have allotted insufficient bits for the results.

One arrives at a quantity referred to as the Kullback information (KI) [10]:

$$KI = +\sum_i^N prob(i) \log_2 \left(\frac{prob(i)}{q(i)} \right)_i$$

$$= \frac{+1}{\log_e(2)} \sum_i^N prob(i) \log_e \left(\frac{prob(i)}{q(i)} \right) \qquad (2.27)$$

$$\approx \frac{+1}{0.693} \sum_i^N prob(i) \log_e \left(\frac{prob(i)}{q(i)} \right)$$

TABLE 2.6

Peptide States and Fractional Occurrences, Observed and Anticipated

Index i	State	prob (i)	$q_i^{(1)}$	$q_i^{(2)}$	$q_i^{(3)}$
1	GGG	0.3000	0.1500	0.2500	0.0020
2	GGV	0.0500	0.1300	0.1000	0.3500
3	GVG	0.0800	0.0600	0.0300	0.4600
4	GVV	0.2000	0.1800	0.2050	0.0020
5	VGG	0.0100	0.1600	0.0050	0.1780
6	VGV	0.2500	0.1700	0.2300	0.0050
7	VVG	0.0200	0.0400	0.0700	0.0020
8	VVV	0.0900	0.1100	0.1100	0.0010

KI quantifies the assessment—here in bits—for anticipating a set of probabilities $\{prob(i)\}$, erroneously or not, according to another probability set $\{q(i)\}$. There is zero penalty (assessment) for correct anticipation: if $prob(i) = q(i)$, then $prob(i)/q(i)$ = 1, and all the logarithm terms in the Equation (2.27) summation cancel to zero. The positive sign emphasizes an important distinction between the formulae used for computing I and KI.

Table 2.6 revisits the data of Table 2.5 along with three examples of anticipated probabilities. Figure 2.4 then follows the lead of Figure 2.3 by presenting one of the scenarios (regarding $q_i^{(1)}$) in picture form. The message is that in arriving at KI, the logarithm arguments $prob(i)/q(i)$ necessarily exceed zero. However, the individual terms in Equation (2.27) can be negative, zero, or positive as marked by the open

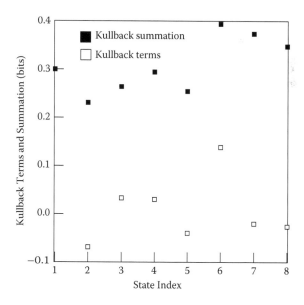

FIGURE 2.4 Kullback terms and summation. The plot is based on anticipated and observed probability data, $q_i^{(1)}$ and $prob(i)$ of Table 2.6.

squares of Figure 2.4. The sum of terms (filled squares) rises and falls but ultimately proves positive or at least never less than zero. The equality with zero only applies when $\{q(i)\}$ *perfectly* matches $\{prob(i)\}$. Clearly, the bit penalties would be greatest if the chemist anticipated the probability set according to the $q_i^{(3)}$ column in Table 2.6. The chemist expected several states—GGG, VGV, VVG, and VVV—to manifest much too infrequently. This led him or her to severely underpurchase the code needed for the state labeling. Note as well that if the chemist had completely missed the boat and expected a probability of zero for any particular state, the bit penalty would have been infinite.

The Kullback information handily applies to situations involving joint probability. The frequent applications entail the mutual information (MI) obtained via summations

$$MI = +\sum_{i,j} prob(i,j) \log_2 \left(\frac{prob(i,j)}{prob(i) \cdot prob(j)} \right)$$

$$= \frac{+1}{\log_e (2)} \sum_{i,j} prob(i,j) \log_e \left(\frac{prob(i,j)}{prob(i) \cdot prob(j)} \right) \qquad (2.28)$$

$$\approx \frac{+1}{0.693} \sum_{i,j} prob(i,j) \log_e \left(\frac{prob(i,j)}{prob(i) \cdot prob(j)} \right)$$

MI has properties that parallel KI. It follows from a sum of weighted logarithm terms that is ultimately positive or at least not less than zero. This is in spite of a mixture of positive and negative terms in the summation. In the example that closed Section 2.3, one considered the joint states formed by the N-terminal unit and the two rightmost amino acids. In the absence of any bias exercised by the cell, $prob(i,j)$ would always equate with $prob(i) \times prob(j) = (1/2) \times (1/4) = 1/8$. One would gauge MI crudely via Equation (2.28) using 1/8 as the denominator in the log arguments and joint probabilities drawn from Table 2.5; in this way MI is estimated to be approximately 0.51 bits. A more involved calculation springs from testing for the independence of the i and j states. In this case,

$$prob(i = 1) = 0.3000 + 0.05000 + 0.08000 + 0.2000 = 0.6300 \qquad (2.29)$$

corresponding to G being observed as the N-terminal unit. By contrast,

$$prob(j = 1) = 0.3000 + 0.01000 = 0.3100 \qquad (2.30)$$

corresponding to GG being the rightmost state. The remainder of the calculation is left as an exercise. That MI exceeds zero means that N-terminal and rightmost units are correlated, although for reasons the chemist would still need to explore. The mutual information shows that the chemist, upon learning the identity of one state, is provided knowledge about the other.

2.5   CONNECTING PROBABILITY, INFORMATION, AND PHYSICAL PROPERTIES

Sections 2.1 through 2.4 addressed states with digital labels attached: 010, 110, VGG, GGV, and so forth. The states of a system, however, manifest abundant physical properties. It is usually through these that a state is identified and the correct label affixed. Information is an unusual quantity because it derives from a weighted average of probability-based terms. The chemist, however, has much more experience measuring molecular weights, densities, and so forth. This section examines physical properties in relation to information.

To begin, any quantitative property tied to a probability set is said to be a random variable. In the traditional notation, the type of property is distinguished from the possible numerical values by capital and small letters; X and x are the usual symbols of choice. For example, X could represent the mass density of a sample; $x = 0.784$ grams per cubic centimeter would then be one of the possible values. Unlike for variables in algebraic and differential equations, there is no rule—at least one apparent to the chemist—which pins down one realization of X over another, hence the designation *random*. The probabilities allied with x nonetheless accommodate graphing and computational techniques. In some situations, X is limited to a finite number of possibilities. Oddly, its statistical nature is often approximated by a continuous function that specifies an infinitude of values. To set the stage, one needs to examine how random variables operate.

When the chemist views states solely in digital terms, the questions and answers are straightforward. What is the probability of observing a tripeptide with sequence VGV? The cell that synthesized the Table 2.5 molecules predicates an answer of 0.2500. When it comes to random variables, however, the questions require modification. For example, the average mass of VGV is computed as 273.3 grams per mole; the same holds for GVV and VGG. Yet it is incorrect for the chemist to inquire what is the probability of observing a peptide manufactured by the cell with mass 273.3 grams per mole. This is because none of the VGV, GVV, and VGG possibilities demonstrate precisely this mass. Such is the case given the isotope combinations of the atoms that compose the molecules. It is instead accurate for the chemist to seek the probability of observing a molecular mass over a specified range. For example, what is the probability of observing a peptide with mass somewhere in a window bounded by 270 and 276 grams per mole? Here a substantial portion of the VVG, VGV, and GVV molecules fit quite nicely. If the window size is decreased, the portion diminishes. If the window is shut and reduced to zero, no member of the population is able to squeeze through.

One arrives at the probability density function $f_X(x)$ for the random variable X having possible values x:

$$f_X(x)\Delta x = \textit{probability of observing x in the range bounded by x and } x + \Delta x \quad (2.31)$$

Key ideas follow straightaway. Probability values are dimensionless: 0.250, 0.075, 10^{-6}, and so forth. Thus the density function $f_X(x)$ must have units of $1/x$, for example, moles per gram. If X referred to the mass density of a sample, $f_X(x)$ would have units of cubic centimeters per gram or an equivalent. Second, probability values are zero, fractional, or at most one. Thus,

$$f_X(x)\Delta x \le 1 \tag{2.32}$$

while the sum of all the probability values equals 1:

$$\sum f_X(x)\Delta x = 1 \tag{2.33}$$

Just as important, since Δx equates with the window size, $f_X(x)\Delta x \to 0$ as $\Delta x \to 0$.

A critical result obtains from partial summations of $f_X(x)\Delta x$. These fall short of one and introduce the probability distribution function $F_X(x \le y)$:

$$F_X(x \le y) = \sum_{x \le y} f_X(x)\Delta x \tag{2.34}$$

$F_X(x \le y)$ measures the probability of observing the random variable X with any value x so long as it is not larger than y. Fundamentals operate here as well. In the most general case:

$$F_X(x) = 0 \quad \text{at} \quad x = -\infty \tag{2.35}$$

More realistically, Equation (2.35) is geared to the physical nature of X: molecular weight, density, temperature, and so on: $F_X(x) = 0$ at $x = 0$. In all cases, however, the distribution function increases, or at least stays constant, as y increases. The maximum size of $F_X(x)$ is clearly 1, the sum of all the normalized probability values. In picture terms, a graph of $F_X(x)$ versus x suggests a curve whose height increases, or stays flat in certain portions, as the x values progress left to right. In some cases, $F_X(x)$ has the appearance of a titration curve.

$f_X(x)$ and $F_X(x)$ are the vehicles for understanding random variables. Both are accessible in venues that pose finite or even an infinite number of possible x. As stated already, data for $f_X(x)$ and $F_X(x)$ can often be modeled by continuous functions, even when the number of states is modest to large. When the chemist identifies which functions apply to a situation, he or she shines light on the system and its statistical nature.

It is worthwhile to demonstrate how $f_X(x)$ and $F_X(x)$ apply to molecular situations. We engage in three thinking exercises involving polypeptides composed of valine (V) and arginine (R) units (cf. Figure 2.2). Let a polypeptide's state be investigated using elementary biochemical and mass spectrometry techniques.

In the first exercise, one considers a 100-unit polypeptide that is 99% V. Let the hypothetical cell place a single R unit randomly in the chain. The possible states are:

$$R_1V_2V_3V_4V_5 \ldots V_{100}$$
$$V_1R_2V_3V_4V_5 \ldots V_{100}$$
$$V_1V_2R_3V_4V_5 \ldots V_{100}$$

$$\cdot$$
$$\cdot$$

$$V_1V_2V_3V_4V_5 \ldots R_{100}$$

It is easy to quantify the information obtained by the chemist inquiring about the system and ascertaining the peptide state. If the peptides are all equally likely,

$$prob(1) = prob(2) = \cdots = prob(100) = \frac{1}{100} \tag{2.36}$$

It follows that

$$I = -\sum_{i=1}^{N=100} prob(i) \cdot \log_2 prob(i)$$

$$= \frac{-1}{100} \log_2\left(\frac{1}{100}\right) - \frac{1}{100} \log_2\left(\frac{1}{100}\right) - \cdots - \frac{1}{100} \log_2\left(\frac{1}{100}\right) \tag{2.37}$$

$$= +1 \cdot \log_2(100)$$

$$\approx \frac{+1}{0.693} \log_e(100) \approx 6.64 \text{ bits}$$

Yet the more critical question is how can a state be ascertained by a physical measurement of a property with attached units and, in turn, $f_X(x)$ and $F_X(x)$? The exercise appeals to the chemical action of trypsin. This is a much-leveraged enzyme that catalyzes peptide cleavage at the carboxyl side of R and K sites [11]. The exercise is one that randomly selects a peptide from a large population followed by trypsin application and isolation of the R- and V-containing product. For extra simplicity, the hypothetical cell never allows K (lysine) as a polypeptide component.

Let X signify the molecular weight of the isolated product with possible x measured in grams per mole. R and V units in free form demonstrate average molecular weights of 174.20 and 117.15 grams per mole, respectively (cf. Table 2.4). It can be shown then that the possible R-containing molecules have approximate molecular weights of:

$$
\begin{array}{ll}
R_1 & \text{174 grams per mole} \\
V_1R_2 & \text{273 grams per mole} \\
V_1V_2R_3 & \text{372 grams per mole} \\
& \quad \cdot \\
& \quad \cdot \\
& \quad \cdot \\
V_1V_2V_3V_4V_5 \cdots R_{100} & \text{9989 grams per mole}
\end{array}
$$

The assignment of a state of the parent randomly selected peptide is obtained by probing the mass of the R-containing descendent over a range of approximately 174 – 10^4 grams per mole. The window size Δx needs to be just less than 100 grams per mole in order to distinguish one molecule—and therefore one parent polypeptide—from another. The window more than accommodates the masses allowed by the isotope combinations. If the chemist observed a fragment with mass ~1956 g/mole, he or she knows at once the state of origin had to have been

VVVVVVVVVVVVVVVVVVVVRVVVVVVVVVVVVVVVVVVVVVVVVVVVVVVVVVVVVVV
VVVVVVVVVVVVVVVVVVVVVVVVVVVVVVVVVVVVVVVVVVVVV

For the above yields

VVVVVVVVVVVVVVVVVVVVR

as the R-containing fragment following the action of trypsin. Note that the chemist is relying on a physical measurement at finite resolution to assign the correct digital label. Matters of probability lie a short step away.

The chemist can well anticipate the statistical structure of the experiment. If the cell randomly places a single R unit, the chemist reasons that

$$f_X(x)\Delta x \approx 0.0100 \qquad (2.38)$$

In turn, $f_X(x) \approx 10^{-4}$ moles per gram over the entire sampling range using the window size Δx of just less than 100 grams per mole.

The probability structure is anticipated by intuition. Computer experiments reinforce the thinking. It is straightforward to construct 100-unit peptides in virtual formats that are 99% V. Likewise, it is simple to place an R unit randomly—using a random number generator—and to compute the molecular weight of the fragment that would be isolated following trypsin application. The R placement and computation are not executed one time only. Rather a large population of peptides and R-containing products must be prepared to realize the probability structure. It was shown in Section 2.1 how the truncated, fractional portion of $(\pi + u)^5$ is virtually unpredictable, given an arbitrary seed $u < 1$. To determine an R placement in a 100-unit peptide, such a fraction can be multiplied by 100 and the result added to 1. The integer portion then points to the site for which to place R. For example, for $u = 0.48902471$, one has

$$(\pi + u)^5 = 630.81556648 \qquad (2.39)$$

$$Integer[100 \times 0.81556648 + 1] = 82 \qquad (2.40)$$

which leads to the peptide

VVVVVVVVVVVVVVVVVVVVVVVVVVVVVVVVVVVVVVVVVVVVVVVVVVVVVVVVVVVV
VVVVVVVVVVVVVVVVVVVVVVVVRVVVVVVVVVVVVVVVVVVV

There is no limit to the virtual molecules that can be prepared in this way. The R placements are sufficiently unpredictable to flesh out the statistical character. Virtual peptides are synthesized more readily by the chemist than real ones.

The upper panel of Figure 2.5 shows a plot of $f_X(x)$ obtained from the computer experiment, whereas the lower panel contains $F_X(x)$. The scatter in the data (open squares) is due to the peculiarities of the random number generator and the finiteness

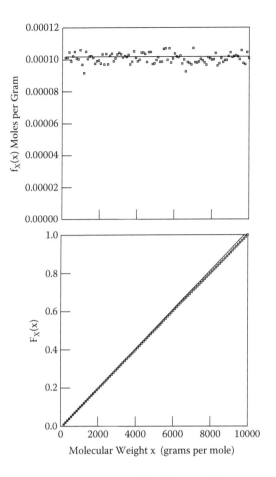

FIGURE 2.5   Statistical structure of V,R peptides. Upper and lower panels show $f_X(x)$ and $F_X(x)$, respectively, corresponding to the uniform distribution. The open squares mark results of the computer exercise described in text. The results of the chemist's intuition have been omitted in the upper panel as they simply follow the horizontal line at height approximately 10^{-4} moles per gram.

of the sample population, in this case 10^5. The linear behavior of $f_X(x)$ and $F_X(x)$ is clear nonetheless and illustrates the major points. Although the possibilities for x were finite in number, their behavior is aptly modeled by the uniform density and probability distribution functions:

$$f_X(x) = \frac{1}{b-a} \tag{2.41}$$

$$F(a \le x \le y) = \frac{1}{b-a} \int_a^y dx' = \frac{1}{b-a} \cdot x' \Big|_a^y = \frac{y-a}{b-a} \tag{2.42}$$

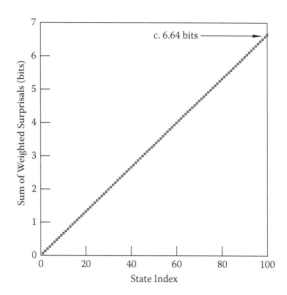

FIGURE 2.6 Sum of weighted surprisals for V,R peptides. The experiment central to the previous figure offers approximately 6.64 bits of Shannon-type information upon establishment of the peptide identity.

For the thinking exercise, best-fit a and b are 174 and 10,087 grams per mole, respectively. Figure 2.5 includes plots (solid line) of the functions in Equations (2.41) and (2.42). As for information, if the chemist labels the states $i = 1, 2, 3, ...$ in order of increasing molecular weight, he or she is a short step from computing the surprisals and weighted summation. The results are shown in Figure 2.6. It is important that the uniform density and probability distribution functions apply to situations where disparate states manifest equal likelihood. Such functions assist in bridging the physical nature of a system with digital labels and information in the quantitative sense.

The second exercise invokes the same cellular machinery, only slightly more error prone. Let the cell produce 100-unit peptides that are almost entirely V. Let the R units be placed randomly at, say, 4% of the sites on average. One of the possible states is

VVVVVVVVRVVVVVVVVVVVVVVVVRVVVVVVVVVVVVVVVVVVVVVVVVVVRV–
VVVVVVVVVVVVVVVVVVVVVRVVVVVVVVVVVVVVVVVVVVVVVVVVVVVVVV

Suppose that the chemist was concerned less about the structure of the molecule as a whole but rather on the distribution of the R-containing sectors. Variations of the previous peptide in which the R units are scattered would seem more likely than molecules in which the R units are bunched together, or are separated by equal numbers of Vs, for example:

VVVVVVVVVVVVVVVVVVVVVVVVRRRRVVVVVVVVVVVVVVVVVVVVVVVVVVVVVV
VVVVVVVVVVVVVVVVVVVVVVVVVVVVVVVVVVVVVVVVVVVVVVVVVV

RVVVVVVVVVVVVVVVVVVVRVVVVVVVVVVVVVVVVVVVVVVRV–
VVVVVVVVVVVVVVVVVVVVVRVVVVVVVVVVVVVVVVVVVVVVVVVVVVVVVVVV

As in the first exercise, trypsin action and measurements of the fragment masses provide a handle on the states and information.

Application of trypsin to the first of the aforementioned peptides yields five fragments:

VVVVVVVVR
VVVVVVVVVVVVVR
VVVVVVVVVVVVVVVVVVVVVVVVR
VVVVVVVVVVVVVVVVVVVR
VVVVVVVVVVVVVVVVVVVVVVVVVVVVV

If the chemist were to probe the R-containing compounds by mass spectrometry, he or she could ascertain the structure of each. How much information is obtained in the typical experiment? The chemist reasons that there is a 4 in 100 chance of any peptide unit being R. There is then a 1 to 0.0400 chance of a unit being V. These considerations link the fragment structure to molecular mass and the likelihood of occurrence:

R	174 grams per mole	$(1 - .04)^0 \cdot 0.0400 = 0.0400$
VR	273 grams per mole	$(1 - .04)^1 \cdot 0.0400 \approx 0.0384$
VVR	372 grams per mole	$(1 - .04)^2 \cdot 0.0400 \approx 0.0369$
VVVR	471 grams per mole	$(1 - .04)^3 \cdot 0.0400 \approx 0.0354$

During the mass probes, the same size window Δx is appropriate as in the first exercise. The chemist's estimates for $f_X(x)$ follow from dividing the probabilities by 99 grams per mole. For example, $f_X(x = 372$ grams/mole$) \approx 0.0369/99$ grams/mole $\approx 3.72 \times 10^{-4}$ moles per gram. Clearly the probability density decreases with the fragment size. Although the V units are assembled by the cell with much greater frequency than R, long sequences of V are not so likely. A computer exercise illustrates the statistical structure and reinforces the chemist's intuition. Here the random number generator directs V at a site if the fractional part of $(\pi + u)^5$ exceeds 0.0400 and injects R otherwise. Upon virtual synthesis of a molecule, a subroutine ascertains the results of trypsin action and computes the masses of the R-containing fragments. When this experiment is carried out many times, say 10^5, the results deliver $f_X(x)$ and other probability quantities by an alternate route.

The chemist's intuition-based $f_X(x)$ is plotted in the upper panel of Figure 2.7 using filled squares. The results of the computer experiment are included as open squares, while the $F_X(x)$ counterparts occupy the lower panel. Figure 2.8 completes matters by showing the sum of weighted surprisals. The agreement between intuition and computer experiment is very good. The consensus is that a query about an R-containing fragment offers approximately 5.86 bits of information. This is less than encountered in the first exercise on account of the reduced uncertainty. In the first exercise, an R unit appeared as 1 of 100 units, whereas in the second it averages 1 in 25. The less the uncertainty, the less information obtained from an experiment.

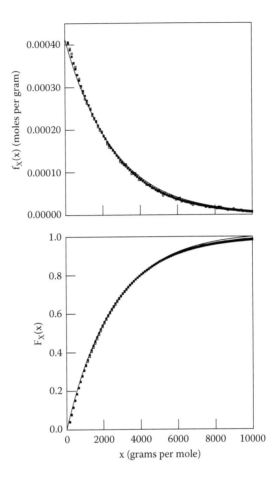

FIGURE 2.7   Statistical structure of more V,R peptides. The upper and lower panels show $f_X(x)$ and $F_X(x)$ corresponding to the exponential distribution. The open squares mark the results of the computer exercise. The filled squares are placed in accordance with the chemist's intuition.

The functional behavior in Figure 2.7 is exponential. This reflects the approximations

$$\exp(-u) \approx 1 - u \tag{2.43}$$

and

$$\exp(-nu) \approx (1 - u)^n \tag{2.44}$$

for $0 \leq u \ll 1$. Accordingly, Figure 2.7 includes plots of the continuous functions

$$f_X(x) = \lambda \exp(-\lambda x) \tag{2.45}$$

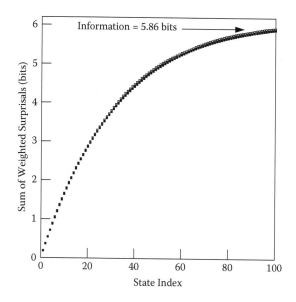

FIGURE 2.8 Sum of weighted surprisals for V,R peptides. The experiment central to the previous figure offers slightly less than 6 bits of information. The open and filled squares are placed, respectively, by the computer exercise and the chemist's intuition.

$$F_X(x \leq y) = \int_0^y dx \cdot \lambda \exp(-\lambda x) \tag{2.46}$$

where λ is a best-fit parameter. Equations (2.45) and (2.46) thereby introduce the exponential distribution. It applies to circumstances that involve random sequences of rare and independent events. The exercise involved the odd placement of an R in a V-rich peptide. Analogous examples can be constructed involving crystalline impurities, chemical side reactions, and radioactive decay.

A third exercise involves cellular chemistry that is unselective altogether. Let the cell produce 100-unit peptides that are mixtures of V and R. Let the R and V units be randomly placed with equal likelihood—the cell is indiscriminant of which unit is placed where. A few of the *many* possible samples are:

VRVRRRVVRRRRRVRRRVRRVRVVRRVVVRVVVVVVVVVVVRRVRVVRVRRVRV –
VVRRRRVRVRRVVVVVRVRVRVVVRRVRRVVVRVVRVVVRRVRRRRR

RVVVVVRRVRVRRRRRRRRVRVRRVRRRVRRRVRVRRVRVVVVVRV –
VVVVVVVVVVVRRRVRVVVVVVRVVVVRRRRRVRRVRVRRRVRVRVVVVVRRVV

VRRVRVRVRVVRRVRVRRVRVVRRVRRRRRRRRRVRRRVVVRVVRVRV –
VVVVVVVVVRVRVRVVVVVVVRVRRVRVVVRVVRVVVRRRRRVRRVRRRRRR

VVRRVRVRVVVVVRRRVRVVRVVVRVVVRRVVVRRRVVRRRRRVRVRVVVVRVRV –
VRRRVRVVRRRRRRVVRRVRVRRRVRRRRVRRVVRRRVRVRRVRVR

VRVRRVVRRRVVRVVRVRRVRRRVRRVRVRRVVVVVVRRVRRRVVRVVVRVR –
RRRVVVVVRRRVVVRVVRRRVRRVRVVVVVVVRVRRRRRVRRVRVRRV

The italics on *many* emphasize that there are indeed multiple allowed states: $2^{100} \approx 10^{30}$. Let the chemist not worry about the primary structure details but only about the number of V versus R units. The chemist can then dispense with the trypsin and proceed directly to the mass spectrometry lab. The window size Δx must be reduced to approximately 57 grams per mole, however. Given random mixtures of R and V in 100-unit molecules, how much information is yielded if the V,R content is ascertained?

Intuition and computation work together again. There is only one configuration each for $(V)_{100}$ and $(R)_{100}$ with respective molecular weights of 9,931 and 15,636 grams per mole. Although these are possible molecules made by the cell, they should manifest with the near-vanishing likelihood of $1/2^{100} \approx 10^{-30}$. By contrast, peptides containing V and R in comparable numbers offer many more possibilities, all at identical molecular weight. Observing these states should be far more likely. The probabilities can be intuited by comparing the number of possible configurations with the total number of allowed states. A polypeptide composed of N number of V units contains $(100 - N)$ R units. The probability $prob(N)$ is quantified by

$$prob(N) = \frac{\left[\frac{100!}{N! \cdot (100-N)!} \right]}{2^{100}} \tag{2.47}$$

for $0 \leq N \leq 100$. The molecular weight of the corresponding peptide, while tedious to compute, arrives by consideration of the V and R weights and linkages as in Figure 2.2. More critical is that the factorial expressions become very large very fast: $6! = 720$ while $(2 \times 6)!$ is 6.65×10^5 times larger. Thus, for almost all the possible cases, the probabilities must be established with the help of Stirling's approximation [6,12]:

$$n! \approx (2\pi n)^{1/2} n^n \exp(-n) \tag{2.48}$$

or in logarithm form

$$\log_e (n!) \approx (1/2) \cdot \log_e (2\pi n) + n \log_e (n) - n \tag{2.49}$$

A computer exercise complements the calculations. Here the random number generator directs V to a site if the fractional portion of $(\pi + u)^5$ exceeds 0.50000 and R otherwise given arbitrary initial $u < 1$. Following virtual synthesis, a subroutine computes the molecular weight of the 100-unit peptide. This experiment is carried out multiple times so as to establish the statistics. Results are illustrated in Figure 2.9. The intuitive estimates based on Equations (2.47) through (2.49) are included. The results for $F_X(x)$ are shown in the lower panel. Figure 2.10 completes the story by showing the sum of weighted surprisals. The information proves to be approximately 4.37 bits. As anticipated, this value is the lowest of the three exercises. This is because the molecular mass measurements do not say as much about the peptide primary structure, only the number of V's versus R's. When the chemist seeks less information, less information is the result. Note one subtle feature. If the chemist is merely concerned about the number of V versus R—not their precise sequence—then there are 100 possible states. Only 40 or so are visited in the typical computer

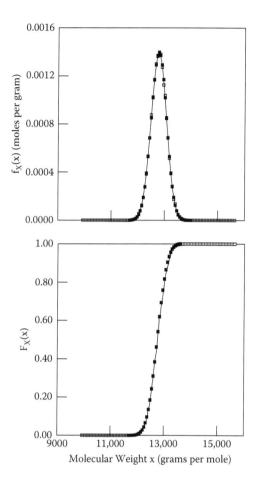

FIGURE 2.9 Statistical structure of random V,R peptides. The upper and lower panels show $f_X(x)$ and $F_X(x)$ corresponding to the normal distribution. The open squares mark the results of the computer exercise while the filled squares are placed by the chemist's intuition. The alignment is near perfect.

experiment; however, 60% of the states are so improbable that they contribute essentially zero weight in the surprisal sum.

Figures 2.9 and 2.10 portray normal or Gaussian behavior. Figure 2.9 thus includes plots of the functions

$$f_X(x) = \frac{1}{\sqrt{2\pi\sigma^2}} \cdot \exp\left[\frac{-(x - x_o)^2}{2\sigma^2}\right] \tag{2.50}$$

$$F_X(x \le y) = \frac{1}{\sqrt{2\pi\sigma^2}} \cdot \int_0^y dx \cdot \exp\left[\frac{-(x - x_o)^2}{2\sigma^2}\right] \tag{2.51}$$

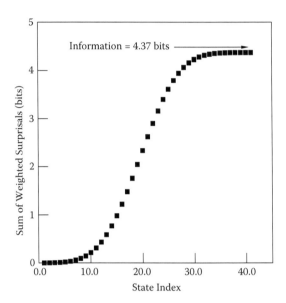

FIGURE 2.10   Sum of weighted surprisals for random V,R peptides. The open squares mark the computational results of the previous figure and the filled squares are placed by the chemist's intuition. The open squares are barely apparent because the alignment is near perfect for the two approaches.

where x_0 and σ are best-fit parameters—their formal significance is addressed in one of the exercises. Note the near-perfect agreement between the continuous functions, the chemist's intuition, and the computational results. The normal distribution applies to multitudinous situations—the adjective *normal* is more than justified. Loosely speaking, Equations (2.50) and (2.51) apply to venues in which a large number of seemingly unrelated variables determine the experimental outcome.

2.6   PROBABILITY DISTRIBUTION TOOLS

Probability functions provide full-length stories about states and likelihoods. Moments and cumulants furnish thumbnail sketches. For a random variable X having a finite number of possible x values, the moments are obtained by weighted summations; the cumulants follow via slightly more complicated formulae. The nth moment is given by:

$$\langle x^n \rangle = \sum x^n f_X(x) \Delta x \tag{2.52}$$

The first moment is then simply the average or expectation of x. Strictly speaking, $\langle x \rangle$ need not be one of the members of the sample population. For example, golfers shoot integer scores with noninteger averages.

The first cumulant K_1 also equates simply with $\langle x \rangle$. The second K_2 is given by:

$$K_2 = \langle x^2 \rangle - \langle x \rangle^2 \tag{2.53}$$

This quantity is tied to the spread of the probability distribution and is referred to as the variance σ^2. The square root of the variance is encountered early and often in science education and is named the standard deviation (σ).

The third cumulant K_3 connects with the symmetry—or more typically the lack of it—in a distribution:

$$K_3 = \langle x^3 \rangle - 3 \cdot \langle x \rangle \cdot \langle x^2 \rangle + 2 \cdot \langle x \rangle^3 \tag{2.54}$$

When continuous functions model the distributions, the moments arrive via integration:

$$\langle x^n \rangle = \int dx \cdot x^n f(x) \tag{2.55}$$

There is an integral to compute for each moment while N integrals are pieced together to supply the Nth cumulant. It is important that the moments can arrive by another route that entails a single integral and an intriguing function. This function can be differentiated sequentially to obtain any moment; it is the moment generating function for the random variable X:

$$M_X(t) = \int dx \cdot \exp[tx] \cdot f(x) \tag{2.56}$$

Comparison of Equations (2.55) and (2.56) shows $M_X(t)$ to be the expectation of $\exp[tx]$. It is interesting to see how such a function is able to furnish all the moments one by one. We first note the Taylor series of $\exp[tx]$ to be:

$$\exp[tx] = \frac{\exp[tx]_{t=0}}{0!} \cdot t^0 + \frac{x^1 \exp[tx]_{t=0}}{1!} \cdot t^1 + \frac{x^2 \exp[tx]_{t=0}}{2!} \cdot t^2 + \frac{x^3 \exp[tx]_{t=0}}{3!} \cdot t^3 + \cdots$$

$$= 1 + \frac{x^1 t^1}{1!} + \frac{x^2 t^2}{2!} + \frac{x^3 t^3}{3!} + \cdots \tag{2.57}$$

As a consequence, $M_X(t)$ equates with

$$M_X(t) = \int dx \left[1 + xt + \frac{x^2 t^2}{2!} + \frac{x^3 t^3}{3!} \cdots \right] \cdot f(x)$$

$$= \int dx \cdot f(x) + t \cdot \int dx \cdot xf(x) + \frac{t^2}{2!} \cdot \int dx \cdot x^2 f(x) + \frac{t^3}{3!} \cdot \int dx \cdot x^3 f(x) + \cdots$$

$$= 1 + t \cdot \int dx \cdot xf(x) + \frac{t^2}{2!} \cdot \int dx \cdot x^2 f(x) + \frac{t^3}{3!} \cdot \int dx \cdot x^3 f(x) + \cdots$$

$$= 1 + t \cdot \langle x \rangle + \frac{t^2}{2!} \cdot \langle x^2 \rangle + \frac{t^3}{3!} \cdot \langle x^3 \rangle + \cdots \tag{2.58}$$

A Taylor series of *any* function $g(t)$ around $t = 0$ is obtained from successive differentiation:

$$g(t) = \frac{g^{(0)} \cdot t^0}{0!} + \frac{1}{1!} \cdot \left(\frac{dg}{dt}\right)_{t=0} \cdot t^1 + \frac{1}{2!} \cdot \left(\frac{d^2g}{dt^2}\right)_{t=0} \cdot t^2 + \frac{1}{3!} \cdot \left(\frac{d^3g}{dt^3}\right)_{t=0} \cdot t^3 + \cdots \quad (2.59)$$

Substituting $M_X(t)$ for $g(t)$ above just repeats the idea of Equation (2.58). The derivatives, one by one, evaluated in the limit $t \to 0$, form the moments $\langle x^n \rangle$.

The workings of $M_X(t)$ are readily demonstrated via the uniform distribution, Equations (2.41) and (2.42). The recipe for the first moment using Equation (2.55) gives:

$$\langle x \rangle = \frac{1}{b-a} \int_a^b dx \cdot x = \frac{1}{b-a} \left[\frac{x^2}{2}\right]_{x=a}^{x=b}$$

$$= \frac{1}{b-a} \times \frac{1}{2} \times (b^2 - a^2) = \frac{b+a}{2} \quad (2.60)$$

To reach the same destination via moment generating, one assembles $M_X(t)$ as follows:

$$M_X(t) = \frac{1}{b-a} \cdot \int_a^b dx \cdot \exp[tx]$$

$$= \frac{1}{b-a} \cdot \frac{1}{t} \cdot [\exp[tx]]_{x=a}^{x=b} = \frac{1}{b-a} \cdot \frac{1}{t} \cdot [\exp[tb] - \exp[ta]] \quad (2.61)$$

Then the first moment $\langle x \rangle$ will arrive from considering the first derivative:

$$\frac{dM_X(t)}{dt} = \left(\frac{1}{b-a}\right) \cdot \left[\frac{-1}{t^2} \cdot \left(\exp[tb] - \exp[ta]\right) + \frac{1}{t} \cdot \left(b\exp[tb] - a\exp[ta]\right)\right]$$

$$= \left(\frac{1}{b-a}\right) \cdot \left[\frac{-\exp[tb] + \exp[ta] + bt\exp[tb] - at\exp[ta]}{t^2}\right] \quad (2.62)$$

One will need to evaluate Equation (2.62) above in the limit $t \to 0$ by applying the rule of l'Hospital. So, both the numerator and denominator in brackets must be differentiated. The operations yield:

$$\left(\frac{1}{b-a}\right) \cdot \left[\frac{-b\exp[tb] + a\exp[ta] + b\exp[tb] + b^2t\exp[tb] - a\exp[ta] - a^2t\exp[ta]}{2t}\right]$$

$$= \left(\frac{1}{b-a}\right) \cdot \left[\left(\frac{-b\exp[tb] + a\exp[ta] + b\exp[tb] - a\exp[ta]}{2t}\right)\right.$$

$$\left. + \left(\frac{b^2\exp[tb] - a^2\exp[ta]}{2}\right)\right] \quad (2.63)$$

which, after cancellations, contribute:

$$\frac{dM_X(t)}{dt} = \left(\frac{1}{b-a}\right) \cdot \left[\frac{b^2 \exp[tb] - a^2 \exp[ta]}{2}\right] \tag{2.64}$$

Equation (2.64), evaluated in the limit $t \to 0$, matches $\langle x \rangle$ given in Equation (2.60). Using $M_X(t)$ to obtain the second moment of the uniform distribution is left as an exercise.

The major points of this chapter are as follows.

1. Three types of information in the quantitative sense were illustrated: Shannon, Kullback, and mutual. They are not the only ones, as information was defined in the 1920s, first by R. A. Fisher [13] then a few years later by R. V. L. Hartley [14]. The 1920s saw the birth of the information sciences due to the technological advances in electronic communication. Applications of the Fisher information lie beyond the scope of this book. The Shannon approach is closely related to Hartley's and is applied in subsequent chapters.
2. How information plays multiple roles was discussed. It equates with the code amounts needed for labeling the states of a system. In turn, it connects with the system's diversity and complexity. A system that requires 160 bits for its state labeling offers far greater message possibilities—and thus diversity—than one needing only 5 bits, for example:

```
0110001111101100010101111000010010110110111111011
1100011111101001101101010101001101000101001011010 0
1000110101101010001011110100000100010111001010001 0
0101101111
```

as opposed to

```
10011
```

By the same metric, a system capable of the 160 bit message is significantly more complicated than one limited to 5 bit communication. If a system expresses one state only, it offers zero information because the diversity and complexity of the messages are absent.

It should also be apparent how information reflects a control capacity. Upon coupling to the environment, a system transmitting a 160 bit message can dictate—or at least influence—a greater number of decisions, compared with a 5 bit counterpart. Decisions impacting the environment invariably entail the transfer of work and heat. Thus, information venues present energy considerations concerning cost and dispersal. The chemist was able to trap information about the polypeptides only by paying a price of work and dissipated heat. Information is physical and chemical; it does not arrive free of charge or independent of work and heat.

3. Links between information and properties such as molecular mass were shown. In each case, the amount of information depends on the nature of states queried and the measurement resolution. If the chemist had narrowed the mass window Δx during the peptide experiments, more information would have obtained due to isotope effects. If he or she applied a mass resolution window of $\Delta x = 20{,}000$ grams/mole, then zero bits would have obtained.

2.7   SOURCES AND FURTHER READING

Information is thoroughly linked to the probability sciences. The author has found several works most instructive over the years. The mathematical probability books by Birnbaum [5], Uspensky [15], Karlin and Taylor [16], along with the Mark Kac lectures are stellar [17]. The book *Randomness* by D. J. Bennett presents a fascinating and wholly accessible approach to probability ideas [18].

Concerning information theory per se, the texts by McEliece [19] and Ash [20] are recommended. The books by Morowitz are especially enlightening and stimulating for students of both chemistry and biology [21,22]. *Science and Information Theory* by Brillouin is more advanced but is indispensible for applications ranging from language to the physical sciences [23]. Wiener's *Cybernetics* includes extended discourse on information, its physical measure and significance [24]. In recent years, the National Research Council has sponsored studies of probability, information, and algorithms. Their report includes a chapter on the generation and significance of random numbers [25]. It should also be mentioned that the statistics of nitration—covered in second-semester organic chemistry—have been thoroughly investigated. One looks to the *Chemical Reviews* article by Ferguson for a complete presentation [26]. Last, the examples of this chapter featured polypeptides. These and their protein counterparts lie at the center of bioinformatics and related fields. The reader will profit from the presentation by Jurisca and Wigle [27]. Chapter 3 of their text, in particular, addresses the mass spectrometry aspects of proteins.

2.8   SUGGESTED EXERCISES

The student should repeat several exercises illustrated in this chapter via the figures and tables. Some computer programming skills and access to a spreadsheet will assist greatly. The same statements apply to exercises of the remaining chapters.

2.1   Information can be reported using bits or nits as units, depending on the logarithm base. There is a third option: the use of base-10 logarithms leads to information measured in Hartleys. (a) How many bits correspond to 8.50 Hartleys? (b) How many Hartleys correspond to 5.30 nits? (c) Invent a unit name for base-18 logarithms. What multiplication factors enable the conversion to bits, nits, and Hartleys?

2.2   (a) Using integral calculus, derive an expression for the second moment of the uniform distribution. (b) Obtain the identical expression by differentiation of $M_X(t)$.

2.3   (a) Using integral calculus, derive expressions for the first, second, and third moments of the standard normal distribution. (b) Do likewise via differentiation of $M_X(t)$.

2.4   (a) How effective is Stirling's approximation as presented in Equation (2.48)? Construct a graph that shows $N!$ versus N in exact terms and by approximation. (b) A more complete version of Stirling's approximation is given by:

$$\log_e(n!) \approx \left(n + \frac{1}{2}\right) \cdot \log_e(n) - n + \log\left(\sqrt{2\pi}\right) + \frac{1}{6 \cdot 2 \cdot 1} \times \frac{1}{n} - \frac{1}{30 \cdot 4 \cdot 3}$$
$$\times \frac{1}{n^3} + \frac{1}{42 \cdot 6 \cdot 5} \times \frac{1}{n^5}$$

Do the extra terms improve matters? Please discuss. (c) Can Equation (2.48) be simplified further for large N? Please discuss.

2.5   Repeat the computer exercise resulting in Figures 2.7 and 2.8. Do the summations of weighted surprisals have to mirror the probability distribution functions? Please discuss.

2.6   The amino acids of Table 2.4 having nonpolar R groups are A, V, L, I, P, F, W, and M. (a) How many 100-unit peptides restricted to these building blocks are allowed? (b) How many bits of information are obtained by the chemist upon learning the sequence of a 100-unit random peptide confined to A, V, L, I, P, F, W, and M?

2.7   The amino acids with uncharged polar R groups are G, S, T, C, Y, N, and Q. Let a large population of 100-unit peptides be prepared using these components chosen at random. (a) Construct a plot of the probability density function based on the peptide molecular weight. (b) Do likewise for the probability distribution function. Do the plots match expectations formed prior to the exercise?

2.8   Repeat Exercise 2.7 without restrictions placed on the amino acids.

2.9   Let a robot synthesize a 100-unit peptide via random selection of the standard 20 amino acids. If a chemist learns from experiment answers to the following questions, how many bits of information are obtained? (a) Is H (histidine) the N-terminal unit? (b) Is H the C-terminal unit? (c) Does the peptide contain two H residues? (d) Does the peptide contain three H residues? (e) Is H absent in the peptide? (f) Does the peptide contain the sequence HVLGA?

2.10 An integer is either prime or composite. Obtain or construct a table of prime integers less than 5000. (a) Let an integer no higher than 1000 be selected at random. How many bits of information are acquired if the status—prime or composite—is determined? (b) Address the same question for a random integer no higher than 5000. (c) Compare and discuss the answers to (a) and (b).

2.10 Look up the formula diagram for cholesterol. (a) How many optical isomers are allowed by the chiral centers? (b) Are the isomers antici- pated in nature with equal likelihood in nature? Please discuss.

2.11 Consider the sequence isomers of lysozyme molecule. This protein has (N- to C-terminal) primary structure:

KVFERCELARTLKRLGMDGYRGISLANWMCLAKWESGYNTRAT-
NYNAGDRSTDYGIFQINSRYWCNDGKTPGAVNACHLSCSALLQD-
NIADAVACAKRVVRDPQGIRAWVAWRNRCQNRDVRQYVQGCGV

(a) How many sequence isomers are allowed? (b) Examine the pairs of nearest-neighbor units: KV, VF, FE, ER, ... Given the occurrence fre- quencies, how many bits of mutual information are expressed in the pair states? Prior to the computation, should the chemist anticipate zero bits of mutual information?

2.12 Let a robot synthesize variants of lysozyme by substituting amino acids at individual sites. (a) Let a chemist know that a single-site sub- stitution has been effected at random. The chemist inquires whether (or not) the sequence is given by:

KVFERCELARTLKRLGMDGYRGISLA**H**WMCLAKWESGYNTRAT-
NYNAGDRSTDYGIFQINSRYWCNDGKTPGAVNACHLSCSALLQD-
NIADAVACAKRVVRDPQGIRAWVAWRNRCQNRDVRQYVQGCGV

How many bits of information are obtained upon learning the answer to the yes–no question? The substitution has been indicated in boldface. (b) In a different experiment, the chemist learns that two sites have been substituted. He or she wonders whether the molecule corresponds to:

KVFERCELARTLKRL**Y**MDGYRGISLANWMCLAKWESGYNTRAT-
NYNAGDRSTDYGIFQINSRYWCNDGKTPGAVNACHLSCSALLQD-
NIADAVACA**R**RVVRDPQGIRAWVAWRNRCQNRDVRQYVQGCGV

How many bits of information are obtained by the answer to the yes–no question? Again, the substitutions have been noted in boldface. Last, con- sider the case of three random site substitutions. The chemist inquires whether the formula is:

KVFERCELAR**H**LKRLGMDGYRGISLANWMCLAKWESGYNTRAT-
NYNAGD**A**STDYGIFQINSRYWCNDGKTPGAVNACHLSCSALLQD-
NIA**G**AVACAKRVVRDPQGIRAWVAWRNRCQNRDVRQYVQGCGV

How much information attaches to the answer?

REFERENCES

[1] Roberts, J. D., Caserio, M. C. 1965. *Basic Principles of Organic Chemistry*, chap. 22, W. A. Benjamin, New York.
[2] Denbigh, K. 1971. *The Principles of Chemical Equilibrium*, chap. 2, Cambridge University Press, Cambridge.
[3] Hill, T. L. 1986. *An Introduction to Statistical Thermodynamics*, chap. 1, Dover, New York.

[4] Shannon, C. E. 1948. A Mathematical Theory of Communication, *Bell Syst. Tech. J.* 27, 379.

[5] Birnbaum, Z. W. 1962. *Introduction to Probability and Mathematical Statistics*, Harper, New York.

[6] Kittel, C., Kroemer, H. 1998. *Thermal Physics*, 2nd ed., chap. 1, W. H. Freeman, New York.

[7] Pauling, L., Wilson, E. B. 1935. *Introduction to Quantum Mechanics with Applications to Chemistry*, chap. 5, McGraw-Hill, New York.

[8] Paša-Tolic, L., Lipton, M. S., eds. 2009. *Mass Spectrometry of Proteins and Peptides. Methods and Protocols*, 2nd ed., Humana Press, Totawa, NJ.

[9] Hempel, J. 2002. An Orientation to Edman Chemistry, in *Modern Protein Chemistry*, Howard, G. C., Brown, W. E., eds., CRC Press, Boca Raton, FL.

[10] Kullback, S. 1997. *Information Theory and Statistics*, Dover, New York.

[11] Smith, B. J. 1994. Enzymatic Methods for Cleaving Proteins, in *Basic Protein and Peptide Protocols*, Walker, J. M., ed., Humana Press, Totawa, NJ.

[12] Wilf, H. S. 1962. *Mathematics for the Physical Sciences*, chap. 4, Dover, New York.

[13] Fisher, R. A. 1925. Theory of Statistical Estimation, *Proc. Camb. Phil. Soc.* 22, 700.

[14] Hartley, R. V. L. 1928. Transmission of Information, *Bell Syst. Tech. J.*, 535.

[15] Uspensky, J. V. 1937. *Introduction to Mathematical Probability*, McGraw-Hill, New York.

[16] Karlin, S., Taylor, H. M. 1975. *A First Course in Stochastic Processes*, 2nd ed., Academic Press, New York.

[17] Iranpour, R., Chacon, P. 1988. *Basic Stochastic Processes: The Mark Kac Lectures*, MacMillan, New York.

[18] Bennett, D. J. 1998. *Randomness*, Harvard University Press, Cambridge, MA.

[19] McEliece, R. J. 1977. *The Theory of Information and Coding: Encyclopedia of Mathematics and Its Applications*, Vol. III, Rota, G. C., ed., Addison-Wesley, Reading, MA.

[20] Ash, R. B. 1990. *Information Theory*, Dover, New York.

[21] Morowitz, H. J. 1970. *Energy for Biologists: An Introduction to Thermodynamics*, Academic Press, New York.

[22] Morowitz, H. J. 1979. *Energy Flow in Biology: Biological Organization as a Problem in Thermal Physics*, Ox Bow Press, Woodbridge, CT.

[23] Brillouin, L. 2004. *Science and Information Theory*, Dover, New York.

[24] Wiener, N. 1961. *Cybernetics*, MIT Press, Cambridge, MA.

[25] *Probability and Algorithms*. 1992. National Research Council Report. National Academy Press, Washington D.C.

[26] Ferguson, L. N. 1952. Orientation of Substitution in the Benzene Nucleus, *Chem. Rev.* 50, 47.

[27] Jurisca, I., Wigle, D. 2006. *Knowledge Discovery in Proteomics*, Chapman & Hall/CRC Press, Boca Raton, FL. Chapter 3 focuses on the mass spectrometry of peptides and proteins.

3 Thermodynamic Infrastructure, States, and Fluctuations

An overview of thermodynamics for elementary systems is presented. We describe the infrastructure for characterizing a system under the very special conditions of equilibrium. It is shown how a maximum entropy state connects with others via fluctuations. The information presented by a system hinges on the statistical structure of the fluctuations.

3.1 INFRASTRUCTURE

Thermodynamics offers numerous pairings:

- Systems *versus* Surroundings
- Work *versus* Heat
- Individual state points *versus* State point loci
- Adiabatic *versus* Diathermal walls
- Closed *versus* Open systems
- Intensive *versus* Extensive properties
- Equilibrium *versus* nonequilibrium conditions
- Finite-time *versus* Infinite-time transformations

There are many more: first- versus second-order phase transitions, state functions versus path-dependent functions, and so forth. However interwoven, the subject can be divided roughly into two parts as presented in Figure 3.1. One part concentrates on the heat and work transferred between a system and its surroundings. The other part attends to the relationships between a system's state variables and functions. There are quite a number of these beginning with temperature (T), pressure (p), and volume (V), as introduced in Chapter 1. If the chemist chooses a quantity such as enthalpy (H), there is quite a story to tell about its relation to other system properties such as compressibility, heat capacity, and so on. Suffice to say that the variables and functions form the infrastructure for thermodynamics under the umbrella of physical laws.

In approaching the subject, one looks first to the state functions centered on potential energy. The most basic of these is the internal energy (U). The nineteenth century

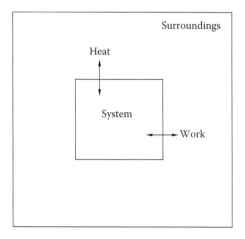

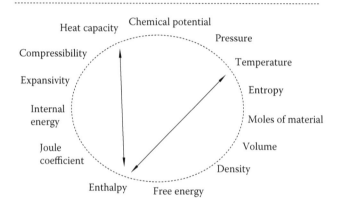

FIGURE 3.1 The dual nature of thermodynamics. Half of the subject (upper panel) focuses on heat and work exchanges between a system and its surroundings. The other half (lower panel) attends to the relationships between state quantities. A property such as enthalpy is related to temperature, heat capacity, and more. The equivalent statement can be made about every quantity in the lower panel.

experiments of James Prescott Joule demonstrated that the energy changes internal to buckets of water were governed by work and heat exchanges. The statement

$$dU = dW_{rec} + dQ_{rec} \qquad (3.1)$$

summarizes the first law of thermodynamics in differential form; dW_{rec} and dQ_{rec} are infinitesimal amounts of work and heat received by a system during some process. In Joule's investigations, the bucket and contents composed the system of interest while everything else acted as the surroundings. The subscripts are important in Equation (3.1) because energy exchanges at once establish two directions. Any work and heat received by a system are identical in magnitude, but opposite in sign, to work and heat lost to the surroundings. One writes:

$$dW_{rec} = -dW_{lost} \tag{3.2}$$

$$dQ_{rec} = -dQ_{lost} \tag{3.3}$$

The work lost by a system is more often described as the work performed or expended. In the terminology used by Joule, the word *duty* is inserted in place of *work*. More important are the landmark ideas of Equations (3.1) through (3.3). The first is that thermodynamics distinguishes two types of energy transfer: work and heat. The former can loosely be described as energy transferred by some orderly arrangement or mechanism. Work demonstrates several guises: mechanical, chemical, electrical, and magnetic. By virtue of the orderliness, there are really no constraints—at least in principle—of converting 10 joules of mechanical work to 10 joules of electrical or vice versa. Hence energy exchanges via work offer innumerable combinations and permutations, in addition to qualifiers. For example, work can be transferred by a system at constant temperature or constant pressure. The infrastructure of thermodynamics draws distinctions between the two situations.

By contrast, heat exchanges are loosely defined as disorderly transfers of energy. Work comes in several flavors, whereas heat is heat whether the source is a Bunsen burner, hotplate, or acid solution mixed with base. The fine print is important just as in work exchanges. Thus, the heat received by a system at constant pressure is not the same as heat received at constant volume.

The second point is that neither dW_{rec} nor dQ_{rec} are *exact* differentials. It is incorrect to write:

$$\int dW_{rec} = W_{rec} \tag{3.4}$$

$$\int dQ_{rec} = Q_{rec} \tag{3.5}$$

because both types of energy transfer depend on the pathway or process details. If one pushes a shopping cart 5 feet down the aisle, or instead directs the cart two laps around the supermarket, back to the starting point, and then 5 feet away, different amounts of work are expended. This is in spite of the same beginning and terminal points. This characteristic of work and heat is often punctuated by combining symbols dW_{rec} and dQ_{rec} with slash marks to yield $đW_{rec}$ and $đQ_{rec}$. In some texts, the inexactness is emphasized by using D in place of d: DW_{rec} and DQ_{rec}. Throughout this book, the path dependence of work and heat will be taken as understood. Slash marks and Ds will not grace the differentials. This does lead to a third point, however.

Although dW_{rec} and dQ_{rec} are inexact differentials, their sum *is* an exact differential by way of dU. It is correct to write:

$$\int (dW_{rec} + dQ_{rec}) = \int dU = \Delta U \tag{3.6}$$

or

$$\int_{initial}^{final} dU = U_{final} - U_{initial} \tag{3.7}$$

where the integration limits and subscripts reference particular states of a system. Another important statement is:

$$\oint dU = 0 \tag{3.8}$$

The integral of Equation (3.8) is unusual in that the initial and final states are taken to be identical. One says that U is a function of state; its changes depend only on the initial and final conditions, not at all on the pathway details. Thermodynamics presents numerous functions of state, and they are not confined to potential energy. V, T, p, mass (m), number of moles (n), and density (ρ) share the list of state functions along with others.

Applications often entail mechanical compression- and chemical-type work, for example, in the operation of an automobile engine. Thus the differential statement for the first law, applied to a one-component system, adopts the form:

$$dU = -pdV + TdS + \mu dn \tag{3.9}$$

where p, V, T, and n have their standard meaning. Entropy (S) and chemical potential (μ) share the stage in Equation (3.9). Note that the three terms to the right of the equal sign correspond to two terms in Equation (3.1). This reflects that work can assert more than one mode of energy transfer simultaneously: in Equation (3.9) via the first and third terms. It is the middle term of Equation (3.9), which equates with dQ_{rec}. One should also note that although dV, dS, and dn are exact differentials, each term on the right in Equation (3.9) is generally inexact. Table 3.1 offers a scorecard for keeping track of the major players or building blocks in thermodynamics. Listed are variables and functions, their extensive or intensive status, and SI units. The extension of Equation (3.9) to systems that host two components, for example, argon and neon, is straightforward:

$$dU = -pdV + TdS + \mu_1 dn_1 + \mu_2 dn_2 \tag{3.10}$$

The pattern is apparent: there is a $\mu_i dn_i$ term for each ith component. For a system hosting κ number of components:

$$dU = -pdV + TdS + \sum_{i=1}^{\kappa} \mu_i dn_i \tag{3.11}$$

Equation (3.11) states that multiple chemical work terms impact the energy exchanges of multicomponent systems.

TABLE 3.1

Building Blocks of Thermodynamics

Variable and Status	Common Symbol	SI (MKS) Unit
Temperature (Intensive)	T	Kelvin
Pressure (Intensive)	p	pascals
Number of moles (Extensive)	n	moles
Mass density (Intensive)	ρ	kilograms/meter^3
Entropy (Extensive)	S	joules/Kelvin
Internal energy (Extensive)	U	joules
Volume (Extensive)	V	meters^3
Isothermal compressibility (Intensive)	β_T	pascals^{-1}
Isentropic Compressibility (Intensive)	β_S	pascals^{-1}
Thermal expansivity (Intensive)	α_p	Kelvin^{-1}
Helmholtz free energy (Extensive)	A	joules
Gibbs free energy (Extensive)	G	joules
Enthalpy (Extensive)	H	joules
Chemical potential (Intensive)	μ	joules/mole

It is the second term on the right of Equation (3.9) that points to the second law of thermodynamics. dS is an exact differential and originates from applying $1/T$ as an integrating factor for dQ_{rec}:

$$dS = \frac{1}{T} \times dQ_{rec} \qquad (3.12)$$

In other words, dQ_{rec} *becomes* an exact differential when it is multiplied by inverse absolute temperature. Note that Equation (3.12) holds strictly for *reversible* changes: the system must never stray from the special condition of equilibrium. If irreversibilities are incurred, then regardless of their origin, $\frac{dQ_{rec}}{T}$ provides only a lower limit for dS. One writes:

$$dS \geq \frac{1}{T} \times dQ_{rec} \qquad (3.13)$$

with equality restricted to equilibrium conditions. Equation (3.13) offers one of several statements of the second law where energy transfer via heat plays a role. A system can be surrounded by thermally insulating walls that preclude such transfer. When such walls are rigid and in place, the conditions are referred to as adiabatic or isentropic. The vacuum jackets of Dewar vessels and thermos bottles offer good approximations of adiabatic walls. For the contents of a Dewar vessel:

$$dS \approx 0 \qquad (3.14)$$

whereby

$$dU \approx -pdV + 0 + \mu dn$$
$$\approx dW_{rec}^{(adiabatic)} \qquad (3.15)$$

Equation (3.15) can be interpreted as the differential statement of the first law under atypical circumstances.

Because dU is an exact differential, it is an explicit function of V, S, and n. This is another lesson of Equation (3.9). Moreover, these three variables all happen to be extensive. For the single component system, one writes:

$$dU = dU(V,S,n) = -pdV + TdS + \mu dn$$
$$= \left(\frac{\partial U}{\partial V}\right)_{S,n} dV + \left(\frac{\partial U}{\partial S}\right)_{V,n} dS + \left(\frac{\partial U}{\partial n}\right)_{V,S} dn \qquad (3.16)$$

For κ-component systems, the extension of Equation (3.16) is cumbersome given the subscript details:

$$dU = dU(V,S,n_1,n_2,\ldots n_\kappa)$$
$$= \left(\frac{\partial U}{\partial V}\right)_{S,n_1,n_2,\ldots n_\kappa} dV + \left(\frac{\partial U}{\partial S}\right)_{V,n_1,n_2,\ldots n_\kappa} dS + \sum_{i=1}^{\kappa} \left(\frac{\partial U}{\partial n_i}\right)_{V,S,n_1,n_2,\ldots n_{j\neq i},\ldots n_\kappa} dn_i \qquad (3.17)$$

Two features are nonetheless apparent. The first is that for a $\kappa = 1$ system, one has, following Equation (3.16),

$$U = -pV + TS + \mu n \qquad (3.18)$$

with obvious extension to multicomponent systems. Second is that long-familiar quantities p and T have partial derivative identities:

$$p = -\left(\frac{\partial U}{\partial V}\right)_{S,n} \qquad (3.19)$$

$$T = +\left(\frac{\partial U}{\partial S}\right)_{V,n} \qquad (3.20)$$

So do less-everyday quantities:

$$\mu = + \left(\frac{\partial U}{\partial n} \right)_{V,S} \tag{3.21}$$

It is an attractive feature of thermodynamics that state variables admit differential expressions; this was remarked upon in Chapter 1. In effect, each variable serves as a reaction of the system to a slight perturbation. Note that the subscripts signal the conditions that are held constant during the infinitesimal change of one variable and the motion of another. The significance of Equations (3.19) and (3.20), and (3.21) by extension, cannot be overstated. T and p are encountered daily via thermometers, barometers, and weather reports. Their physical nature runs much deeper, however. T connects with how a system's energy behaves, given miniscule adjustments of the entropy under leak-proof conditions. A parallel statement holds regarding system p. Also notable is that variables related to one another by a derivative of U play conjugate roles. p and V are conjugate to each other; likewise for T and S, μ and n. As Figure 3.1 indicates, there is a story to tell about how each member of the extended family is related to another.

Thermodynamics has no shortage of quantities that are differential in nature. Several are stated as follows:

$$C_V = \left[\frac{dQ_{rec}}{dT} \right]_{V,n} = T \left(\frac{\partial S}{\partial T} \right)_{V,n} \tag{3.22}$$

is the heat capacity for a one-component system at constant volume;

$$C_p = \left[\frac{dQ_{rec}}{dT} \right]_{p,n} = T \left(\frac{\partial S}{\partial T} \right)_{p,n} \tag{3.23}$$

is the heat capacity at constant pressure;

$$\alpha_p = \frac{1}{V} \left(\frac{\partial V}{\partial T} \right)_{p,n} \tag{3.24}$$

is the thermal expansivity (expansion coefficient) at constant pressure; and

$$\beta_T = \frac{-1}{V} \left(\frac{\partial V}{\partial p} \right)_{T,n} \tag{3.25}$$

is the compressibility at constant temperature. Its counterpart

$$\beta_S = \frac{-1}{V} \left(\frac{\partial V}{\partial p} \right)_{S,n} \tag{3.26}$$

applies to isentropic conditions. The negative signs in Equation (3.25) and Equation (3.26) are not throwaway details. Their presence ensures that β_T and β_S are positive quantities for all stable systems.

For the differentials of Equations (3.22) through (3.26), n is held constant. The omission of n from the parentheses subscripts is a frequent practice. This is analogous to the omission of carbon and hydrogen atom symbols from chemical formula diagrams of the line-angle variety; the experienced viewer infers their presence. It is important that C_V, C_p, α_p, β_T, and β_S apply to closed systems. They are all positive for stable systems—the type under the lens of this chapter.

Thermodynamic differentials are not restricted to the first order. Further, since the order of differentiation is immaterial to the outcome, one has that

$$\left[\frac{\partial}{\partial S}\left(\frac{\partial U}{\partial V}\right)_{S,n}\right]_{V,n} = \left[\frac{\partial}{\partial V}\left(\frac{\partial U}{\partial S}\right)_{V,n}\right]_{S,n} \tag{3.27}$$

It follows that

$$-\left(\frac{\partial p}{\partial S}\right)_{V,n} = \left(\frac{\partial T}{\partial V}\right)_{S,n} \tag{3.28}$$

Second derivative identities such as these are referred to as Maxwell identities. Second derivatives that involve the chemical potential are equally valid although encountered infrequently. One example is:

$$\left[\frac{\partial}{\partial n}\left(\frac{\partial U}{\partial V}\right)_{S,n}\right]_{V,S} = \left[\frac{\partial}{\partial V}\left(\frac{\partial U}{\partial n}\right)_{V,S}\right]_{S,n} \tag{3.29}$$

whereby

$$-\left(\frac{\partial p}{\partial n}\right)_{V,S} = \left(\frac{\partial \mu}{\partial V}\right)_{S,n} \tag{3.30}$$

A system's internal energy depends on the component identity, mole amount, and phase; for example, 2.00 moles of helium gas versus 1.00 mole of xenon gas versus 1.50 moles of liquid ammonia. Yet the dependence of U on V and on S is not entirely case specific. One gathers this from the second derivatives linked to functions such as in Equation (3.22). Differentiating U twice with respect to S leads to:

$$\left(\frac{\partial^2 U}{\partial S^2}\right)_{V,n} = \left[\frac{\partial}{\partial S}\left(\frac{\partial U}{\partial S}\right)_{V,n}\right]_{V,n}$$

$$= \left(\frac{\partial T}{\partial S}\right)_{V,n} \tag{3.31}$$

$$= \frac{T}{C_V}$$

Since heat capacity and absolute temperature are positive for equilibrium systems, U is mandatorily a *concave upward* function of S. Plots of U versus S demonstrate a signature shape with no exceptions. In the same vein,

$$\left(\frac{\partial^2 U}{\partial V^2}\right)_{S,n} = \left[\frac{\partial}{\partial V}\left(\frac{\partial U}{\partial V}\right)_{S,n}\right]_{S,n}$$
$$= -\left(\frac{\partial p}{\partial V}\right)_{S,n} \qquad (3.32)$$
$$= \frac{1}{V\beta_S}$$

Compressibilities are positive because systems lose volume during compression. A consequence is that U is also a concave upward function of V; plots of U versus V have signature features, regardless of the component, phase, and mole amount. Figure 3.2 accordingly illustrates the signature behavior of U. The point being made

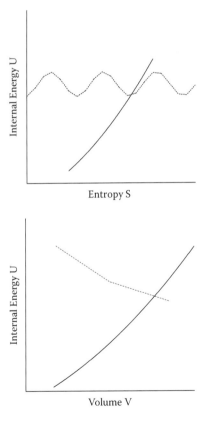

FIGURE 3.2 The signature dependence of U on volume and entropy. U versus V and U versus S mandate concave-upward curves. The curves remain above any and all tangent lines that can be drawn. The dotted curves are examples of functions strictly forbidden for stable systems.

is that U applies to diverse chemical systems, the *numerical* details of which will always be case specific. The stability of a system, however, places formidable restrictions on the dependence of U on V and S.

The internal energy is not the only potential. Others are obtained via Legendre transformations of U, for example:

$$H = U - \left(\frac{\partial U}{\partial V}\right)_{S,n} \cdot V$$

$$= U + pV$$

$$= -pV + TS + \mu n + pV \tag{3.33}$$

$$= TS + \mu n$$

The function H obtained by the transformation depends explicitly on S and n held constant, while the third variable is the "new" one manifest in the derivative. Thus $H = H(p,S,n)$ for a $\kappa = 1$ system and is referred to as the enthalpy. In a like way, other useful potentials are obtained, for example:

$$A = U - \left(\frac{\partial U}{\partial S}\right)_{V,n} \cdot S = U - TS = A(V,T,n)$$

$$= -pV + TS + \mu n + pV - TS \tag{3.34}$$

$$= -pV + \mu n$$

Legendre transformations are not limited to one variable. By differentiating and multiplying twice, one obtains:

$$G = U - \left(\frac{\partial U}{\partial V}\right)_{S,n} \cdot V - \left(\frac{\partial U}{\partial S}\right)_{V,n} \cdot S = U + pV - TS = G(p,T,n)$$

$$= -pV + TS + \mu n + pV - TS \tag{3.35}$$

$$= \mu \cdot n$$

A and G are referred to as the Helmholtz and Gibbs free energy, respectively. Note that terms *enthalpy* and *free energy* allude to heat and work, respectively. The terminology subscribes to the circumstances where the potentials are most often directed. For instance:

$$dH(S,p,n) = TdS - Vdp + \mu dn \tag{3.36}$$

for a single-component system. If the system is closed and the pressure held constant, then

$$[dH]_{p,n} = TdS = [dQ_{rec}]_{p,n} \tag{3.37}$$

For this reason, enthalpy H is a natural fit for situations that entail heat exchanges at constant pressure. These exchanges transpire daily inside and outside chemistry labs, such that H is encountered frequently in thermodynamics. Along similar lines,

$$dA(V,T,n) = -pdV - SdT + \mu dn \tag{3.38}$$

If the system is closed and its temperature held fixed, then

$$[dA]_{T,n} = -pdV = [dW_{rec}]_{T,n} \tag{3.39}$$

The free energy A is thus geared for systems in which work is transferred at constant temperature. It should be noted that no single potential deserves more attention than another. What is true is that some are better suited for certain conditions. For example, G has special stature in chemistry via Equation (3.35). It follows that

$$dG = \mu dn + nd\mu \tag{3.40}$$

G thereby connects with situations in which chemical work is paramount. The applications are without limit and are discussed in first-year chemistry courses and beyond.

U, H, A, and G form the short list, but not the whole list, of potential energy functions. Potentials such as Φ are equally valid and obtain by Legendre transformation through n:

$$\Phi = U - \left(\frac{\partial U}{\partial n}\right)_{V,S} \cdot n = U - \mu n = \Phi(V,S,\mu)$$

$$= -pV + TS + \mu n - \mu n \tag{3.41}$$

$$= -pV + TS$$

However, they find sparse applications in chemistry. It is important to note the symmetry in the potentials and variables. Symmetry properties indeed underpin diagrams such as in Figure 3.3. The layout is referred to as the *thermodynamic square* in most texts. The logistics are that each short-list potential is situated between the variables upon which it depends explicitly. At the same time, the variables that conjugate to each other appear at opposite corners. The diagram omits reference to n as this quantity is fixed in most applications.

The uses of the diagram are multiple and interesting. The diagram helps the chemist to remember which variables are allied with which potential—it serves as another scorecard so to speak. On what variables does A depend explicitly? A glance at Figure 3.3 identifies V, T—and n by default—as the answer. Second, rectangles in the mind's eye help the chemist track how the potentials are related to one another: the Legendre transforms are an imbedded feature. How are U and H related? The chemist places his or her finger at the tail of an arrow and follows the route spelled out inside the dotted rectangle, that

$$pV + U = H \tag{3.42}$$

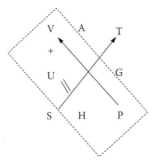

FIGURE 3.3   The thermodynamic square and one application. A Legendre transformation is implicit inside the dotted rectangle. The chemist starts at the tail of the arrow at lower right, imagines the plus and equal signs, and assembles $pV + U = H$.

Equation (3.42) is a simple restatement of Equation (3.33). The chemist can use the diagram to fill in the blanks featuring total differentials and variables, for example:

$$[dU]_n = ??\, dV + ??\, dS \tag{3.43}$$

The question marks refer to the variable multipliers and their signs. The arrows point toward the multipliers as the conjugate variables. This clue yields:

$$[dU]_n = ?\, pdV + ?\, TdS \tag{3.44}$$

Then which way an arrow points asserts the sign. If an arrow points away from the multiplier, the sign is positive, and negative otherwise. One learns from Figure 3.3:

$$[dU]_n = -pdV + (+)TdS \tag{3.45}$$

which is a variant of Equation (3.9).

Using the diagrams to construct first-derivative identities is straightforward. This is shown via Figure 3.4. An imaginary pipe encloses the differentiation of a potential with respect to one variable. The arrow along the stem of the pipe identifies the variable on the opposite side of the equal sign. If the arrow points toward the conjugate variable, the sign is positive and negative otherwise. One has to imagine smoke wafting from the pipe barrel—the upward-pointing arrow in Figure 3.4. The recipient of the smoke is the variable held constant during differentiation while constant n is implicit. For the artistically inclined, there are eight pipe drawings inspired by the diagram.

Last, imaginary triangles access the Maxwell identities. The diagram is written twice and triangles are drawn as in the lower half of Figure 3.4. The center identifies the differentiation variables; the right and left corners host the variables held constant. The arrows keep all the signs correct. When the arrows point in opposite directions, there is a sign inversion. It is left to the reader to construct four sets of triangles and affiliated Maxwell relations. There is one set for each 90-degree rotation and side-by-side rendering of the diagram.

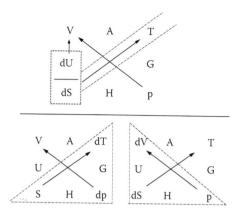

FIGURE 3.4 The thermodynamic square and two more applications. A first-derivative identity is implicit in the dotted pipe of the upper panel. The chemist imagines differential signs in front of U and S. Immediately above U is V, which is held constant during differentiation. The diagonal arrow points to T as in $\left(\frac{\partial U}{\partial S}\right)_{V,n} = T$. In the lower panel, the side-by-side diagram illustrates the second derivative, or Maxwell identity $\left(\frac{\partial T}{\partial p}\right)_{S,n} = \left(\frac{\partial V}{\partial S}\right)_{p,n}$.

3.2 EQUATIONS OF STATE

Potentials, variables, and laws comprise the majority of the thermodynamic infrastructure. The remainder is obtained via equations of state ranging from the rigorous to empirical. The equations vary in their names, history, and familiarity. This section will concentrate on the empirical category for its ease of application. The rigorous category includes infinite series or virial-type equations. These describe systems precisely, although at the cost of weighing an infinite number of terms. The empirical variety affords simplicity via economy.

The first equation of state is the most familiar, namely, for an ideal gas:

$$p = \frac{nRT}{V} \tag{3.46}$$

R is the well-known constant with units of entropy per mole. R first appeared in Equation (2.18) for the entropy of mixing; it is a close relative of Boltzmann's constant, first appearing in Equation (2.19):

$$k_B = \frac{R}{N_{Av}} \tag{3.47}$$

where N_{Av} is the Avogadro number. Table 3.2 lists R and k_B for three different unit systems. Note that since

$$n = \frac{N}{N_{Av}} \tag{3.48}$$

TABLE 3.2

Gas Constant (R), Boltzmann's Constant (k_B), and Units

	MKS	CGS	Non-SI
R	$8.31 \dfrac{joules}{mole - K}$	$8.31 \times 10^7 \dfrac{ergs}{mole - K}$	$0.0821 \dfrac{liter - atmospheres}{mole - K}$
k_B	$1.38 \times 10^{-16} \dfrac{ergs}{K}$	$1.38 \times 10^{-23} \dfrac{joules}{K}$	$1.36 \times 10^{-25} \dfrac{liter - atmospheres}{K}$

where N is the number of gas particles, an alternative to Equation (3.46) is:

$$p = \frac{Nk_B T}{V} \tag{3.49}$$

Although the stockroom supplies only real gases in tanks—argon, carbon dioxide, and so forth—Equations (3.46) and (3.49) apply to every system in the limit of zero density. The compactness of the ideal gas law is impressive as four variables of state are related by simple multiplication and division. The equation wraps Boyle's law, Charles law, and the Avogadro hypothesis into a compact multivariable function.

A second empirical equation of state is named after Clausius:

$$p = \frac{nRT}{V - nb} \tag{3.50}$$

This is the forerunner of the more famous van der Waals equation:

$$p = \frac{nRT}{V - nb} - \frac{an^2}{V^2} \tag{3.51}$$

Equations (3.50) and (3.51) introduce parameters a and b, which depend on the material of interest. The parameters are taken to be independent of T, V, p, and n. Equations (3.50) and (3.51) can be viewed as modest extensions of the ideal gas law. The term nb gauges the volume excluded by the gas atoms or molecules. It reflects that the same space cannot be occupied, at least at the same time, by different parties. The reason is clear in modern day; not so much in the era of Clausius and van der Waals. Negative charges form the periphery of an atom or molecule and repel any neighbors that approach too closely. The consequence is that the volume available to the atom or molecule is always something less than that of an evacuated container. Figure 3.5 addresses matters schematically: the upper portion shows a box of gas molecules and the lower presents a close-up view of a sector indicated by the inset square. Because a real molecule excludes volume, it causes its neighbors to exert a pressure greater than anticipated by the ideal gas law. In turn, this means that for a gas to be truly ideal, its constituent atoms or molecules would have to claim zero space. Instead of filled circles in the lower part of Figure 3.5, one would have to represent geometric

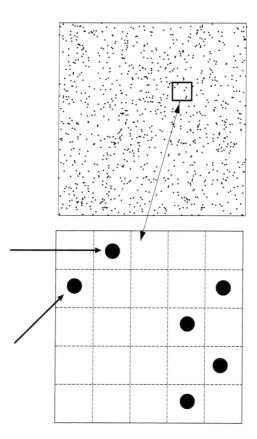

FIGURE 3.5 Gases, excluded volume effects, and probability. The upper and lower diagrams illustrate N particles in volume V on disparate length scales. Two molecules such as indicated by the arrows cannot occupy the same imaginary cells at the same time. The probability of molecules occupying neighboring cells scales in the manner of the second pressure term of the van der Waals equation.

points of zero width. The ideal gas law is one of numerous idealizations in thermodynamics; the adjective *ideal* is well justified.

In early days, b values offered first insights into the sizes of atoms and molecules. One way b can be accessed experimentally is for the chemist to prepare a gas sample and ascertain p, n, T, and V. The chemist follows up by measuring the differential quantity

$$\left[\frac{\Delta p}{\Delta T} \right]_{V,n} \approx \frac{nRT}{V - nb} \tag{3.52}$$

Equation (3.52) can be rearranged to isolate b. The approximation is better than not if Δp and ΔT are kept as small as possible in the experiment. Table 3.3 lists van der Waals a and b for several gases, as compiled in the *Handbook of Chemistry and Physics* [1]. It is interesting to correlate a and b with the complexity and size of a given atom or molecule.

TABLE 3.3

Van der Waals Constants of Assorted Atoms and Molecules

Atom/Molecule	$a\left(\dfrac{\text{meters}^6 \cdot \text{pascals}}{\text{mole}^2}\right)$	$b\left(\dfrac{\text{meters}^3}{\text{mole}}\right)$
Acetic acid	1.776	1.068×10^{-4}
Acetone	1.404	9.94×10^{-5}
Argon	0.1358	3.219×10^{-5}
Ethane	0.5543	6.38×10^{-5}
Helium	0.003446	2.37×10^{-5}
Hydrogen	0.02468	2.661×10^{-5}
Hydrogen chloride	0.3703	4.081×10^{-5}
Hydrogen bromide	0.4495	4.431×10^{-5}
Mercury	0.8173	1.696×10^{-5}
Methane	0.2275	4.278×10^{-5}
Propane	0.8750	8.445×10^{-5}
Propylene	0.8462	8.272×10^{-5}
Sulfur dioxide	0.6781	5.636×10^{-5}
Xenon	0.4235	5.105×10^{-5}
Water	0.5519	3.049×10^{-5}

Source: Data from Weast, R. C., ed. *Handbook of Chemistry and Physics*, Chemical Rubber Co., Cleveland, OH, p. D146, 1972.

The second term on the right in the van der Waals equation arises from additional forces operating in a gas. The negative charges of a molecule do more than repel the neighbors. Rather, their density fluctuations switch on short-range attractive forces. In contrast to excluded volume effects, the forces of attraction diminish the pressure values anticipated by the ideal gas law. The second term on the right in Equation (3.51) scales as the square of the number density (n/V). Importantly, the scaling connects with probability ideas in the close-up view of Figure 3.5. The gas molecules have been imagined as occupying cells of equal volume $\bar{V}$; one molecule fits into one cell and, under low density conditions, has an exceedingly large number λ from which to choose: $V = \lambda \cdot \bar{V}$. For a sample of N molecules, the probability of any particular cell being occupied at a given instant is proportional to $(N/\lambda \cdot \bar{V})$. The probability of any given cell and a neighbor being occupied is proportional to $(N/\lambda \cdot \bar{V}) \times (N/\lambda \cdot \bar{V})$; this assumes more or less independent behavior of the gas particles. There are two points being made here. First, the effects of both repulsive and attractive forces on the gas pressure have a statistical nature and impact. Second, there is a natural bridge linking even the most elementary of systems with probability ideas. One need only look at the gas laws of first-year chemistry courses to see the bridge.

Note as well that the second term of the van der Waals equation scales as *length*$^{-6}$. The interaction energies of mutually induced dipoles scale in a like fashion. In the limit of large V and high T, Equations (3.50) and (3.51) operate more like the ideal gas law. This is consistent with experiments; nonideal behavior is accentuated under high-density, low-temperature conditions. Adherence to the ideal gas law is the trademark of high-temperature, low-density systems.

There are numerous equations of state of the empirical variety. There is the Dieterici equation:

$$p = \frac{nRT}{V - nb} \times \exp\left[\frac{-an}{RTV}\right] \tag{3.53}$$

Like the van der Waals, Equation (3.53) weighs the competing effects of excluded volume and short-range attractive forces using two parameters. The Beatie–Bridgeman equation appeals to three parameters:

$$p = \frac{nRT \cdot (1 - \varepsilon) \cdot (V + nb)}{V^2} - \frac{an^2}{V^2} \tag{3.54}$$

The Bertholet equation of state is:

$$p = \frac{\left\{RT\left[1 + \frac{9}{128}\left(\frac{p}{p_c}\right)\left(\frac{T_c}{T}\right)\left(1 - \frac{6T_c^2}{T^2}\right)\right]\right\}}{V} \tag{3.55}$$

and incorporates the effects of critical pressure and temperature, p_c and T_c. Other equations of state include ones constructed by Redlich, Kwong, and Soave, and by Peng and Robinson [2]. An equation designed by Benedict, Webb, and Rubin employs no fewer than eight parameters [3].

Thermodynamics has logged numerous equations of state over the years of the empirical variety. The common thread is their economy and intuition, where only a few parameters are called upon to address a constellation of forces. As the equations entail multivariable functions, they accommodate the tools of first-year calculus and, in turn, the infrastructure presented in Section 3.1. In their most basic applications, they enable conversions of independent variables into dependent ones. This is the subject of Figure 3.6. The ideal gas and van der Waals equations are represented as input–output devices. The devices accept n, T, V, measured for a gas such as argon and generate p in return. At 200 K, 0.00150 meter^3, 2.00 moles, the van der Waals equation, with the help of Table 3.3 data, offers $p = 2.07 \times 10^6$ pascals, while the ideal gas law delivers $p = 2.22 \times 10^6$ pascals. The values differ because the nonideality is addressed, at least in part, by one device and not the other. The result is that the van der Waals equation better approximates the location of what will be termed the *state point* of the system: the placement of a point in a coordinate plane such as pT. The state point placement is represented schematically in the lower half of Figure 3.6.

Point locations are a universal purpose of equations of state. Yet every application arrives with some fine print that is not altogether obvious. The print indeed imposes qualifiers that the chemist must always keep in mind. The first is that input–output conversions as in Figure 3.6 apply only to systems that are truly at equilibrium. For nonequilibrium venues, equations such as the van der Waals, Dieterici, and so on are

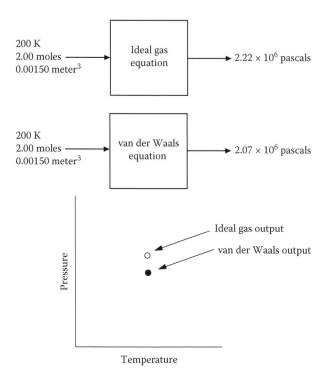

FIGURE 3.6 Equations of state, input and output. The ideal gas and van der Waals equations are portrayed as devices that convert input variables into output. The lower portion depicts the location of state points in the *pT* plane anticipated by the output.

really ineligible for locating state points. Of course, this raises the question of what is meant by an equilibrium as opposed to nonequilibrium system.

The second qualifier is that when an equation of state converts input to output, the latter should be viewed as an *estimate* of an *average value*. That is to say, the probability issues raised in Figure 3.5 are not apparent in the equations taken at face value. Yet their approximation nature extends beyond using only a few parameters to address molecular-level forces and the physical uncertainties surrounding R. Unlike functions encountered in calculus books, for example:

$$f(x,y,z) = \frac{3xy}{z} + 4xy^2 - 10xz + 6 \qquad (3.56)$$

the output of a thermodynamic equation of state does not correspond to a point that is infinitely sharp. In other words, the width of each point in the *pT* plane of Figure 3.6 is not simply a graphing artifact. This subtle feature of chemical systems and equations of state is important and is given more attention in Section 3.3. For now, one aims for a clearer picture of an equilibrium state—the type eligible for study using the ideal gas law, van der Waals equation, and so on.

Equilibrium states have a number of characteristics. For a single-phase system, they are the states devoid of sharp and persistent gradients in the intensive quantities:

temperature, density, chemical potential, and so forth. Equilibrium states are the ones associated with the maximum entropy allowed by the conditions. They are the states that afford a chemist zero work for the taking, unless further constraints are removed. They are the most probable states of a system, given the circumstances. They are the states stable to fluctuations. Equilibrium states have a nature that is interminably robust and restorative. The characteristics do not make a trivial list.

Chapter 2 introduced probability elements via thinking and computer exercises. The nature of an equilibrium state can be grasped the same way. A few examples are illustrated that rely on ideal gases for simplicity. The ideas are borne out just as well using empirical equations of state such as van der Waals. The computations become more involved, however.

Perhaps the optimum way for a chemist to comprehend equilibrium conditions is to imagine a system away from equilibrium in a single idealized respect. One considers the container shown in the top frame of Figure 3.7. Illustrated is a system of twin compartments in thermal contact with one another. Let the adjoining wall be composed of heat conducting material, while the system as a whole is surrounded by a leak-proof adiabatic wall. The compartments are identical in size and contents. Let each house 1.00 mole of neon in a volume of 1.00 meter^3. The exercise begins with the left compartment at 300 K and the right at 400 K. Let there be no other temperature gradients, thus allowing application of the ideal gas equation to each compartment viewed individually. In other words, let the contents of each compartment maintain a state of local equilibrium; let a true and single-value temperature apply at all times to each. The idealizations notwithstanding, it is clear what will happen. Because the dividing wall is thermally conducting, heat will flow from right to left. The transfer is obtained by collisions of the atoms with themselves and the walls, although these need not be addressed in any detail. More important is the denouement of the story. The left-side neon will warm to 350 K while the right will cool to the same temperature. When this condition is in place, thermal equilibrium will apply to the system as a whole.

Let the neon be modeled by the ideal gas law. This is reasonable given the high-temperature and low-density conditions. The internal energy U will then depend only on the gas amounts and the temperature. The first law for a closed system (cf. Equation 3.9) holds that

$$[dU]_n = -pdV + TdS \tag{3.57}$$

Further, elementary kinetic theory contributes the relation:

$$U = \left(\frac{3}{2}\right)nRT \tag{3.58}$$

for an ideal monatomic gas. By linking Equations (3.57) and (3.58) under constant volume conditions, one obtains:

$$[dU]_{V,n} = TdS = \left(\frac{3}{2}\right)nR \cdot dT \tag{3.59}$$

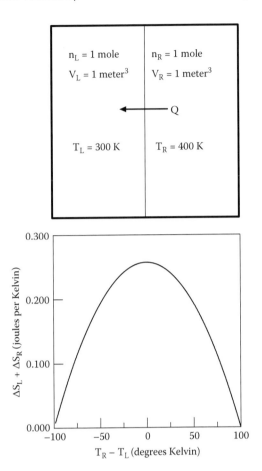

FIGURE 3.7 Composite systems and thermal equilibrium. Illustrated are twin compartments in thermal contact. Heat Q flows from right to left until equilibrium is established. The lower frame shows the total entropy change as a function of the temperature difference between the two sides. The change is maximum when the difference is zero.

whereupon

$$\left(\frac{\partial U}{\partial T}\right)_{V,n} = T\left(\frac{\partial S}{\partial T}\right)_{V,n} = C_V = \left(\frac{3}{2}\right)nR \tag{3.60}$$

The result is

$$[dU]_{V,n} = TdS = C_V dT \tag{3.61}$$

which applies to both left- and right-side neon. Any heat lost by the warmer gas is absorbed by the cooler. Since the heat capacities of the compartments are identical, a diminution of T on the right is answered by an equal rise on the left. Concerning the entropy changes, one has from Equation (3.61) that

$$dS = \frac{1}{T} \cdot C_V dT \tag{3.62}$$

hence

$$S_j - S_{initial} = \int_{T_{initial}}^{T_j} C_V \frac{dT}{T} \tag{3.63}$$

$$= C_V \cdot \log_e \left(\frac{T_j}{T_{initial}} \right)$$

Equation (3.63) is a means for quantifying the entropy change in each compartment as the equilibrium is approached through a succession of j-labeled states. Since one side warms while the other cools, the entropy changes are positive on the left and negative on the right. This offers a truism for equilibration processes. The total entropy change is positive although the change need not be positive everywhere.

The lower panel of Figure 3.7 illustrates the total entropy change for the system as a function of the temperature difference between the left and right sides. It is apparent that the change is greatest when the difference is zero. This is the key result. Thermal equilibrium is obtained not just when T is uniform in a material but also when the entropy has been maximized. But note also the robustness. If too much heat is transferred from right to left, then the total entropy will not be at the maximum value. The effects are fleeting, however, as a new temperature gradient is able to drive the back-transfer of heat. Note further a subtle yet critical feature. The equilibrium state is not singular in nature. Rather, it comprises a set of states in the vicinity of the maximum entropy one. Moreover, when a system demonstrates the maximum possible entropy, a minimum number of variables (e.g., a single temperature) suffice to describe it. When and where the system strays from maximum entropy, more facts and data are needed by the chemist to detail the thermodynamic conditions.

A follow-up exercise looks at the system of Figure 3.8. Featured are twin compartments, this time separated by a diathermal (heat conducting) wall that is free to move on frictionless bearings. The compartments house different amounts of neon gas: 1.00 and 2.00 moles for the left and right, respectively. Both sections are at 300 K with equal initial volumes of 1.00 meter^3. Let no pressure gradients persist in each compartment such that local equilibrium states are in effect—a single pressure value applies to a compartment at any given instant. Clearly the force disparity of the two sides will cause the wall to slide right to left. The initial pressure difference between compartments arrives by the ideal gas law:

$$\Delta p = \frac{(n_R - n_L)RT}{V} \tag{3.64}$$

$$\approx 2.49 \times 10^3 \, pascals$$

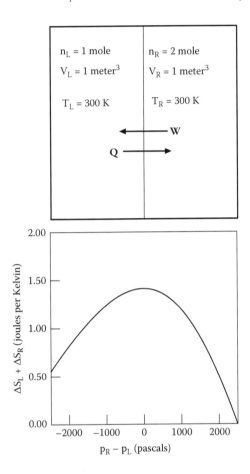

FIGURE 3.8 Composite systems and mechanical equilibrium. The barrier slides right to left in the equilibration process. In turn, work W flows from right to left and heat Q flows in the opposite direction. Heat and work flow until the pressures equalize. The lower frame shows the total entropy change as a function of the pressure difference between the compartments. The change is maximum when the difference is zero.

Just as apparent is that the right-side gas will expand, thereby transferring work to the left. The wall will cease moving when the pressures of the two sides match. Since the center wall is diathermal, any heat of compression felt on the left will flow toward and be absorbed by the right. This ensures the 300 K conditions prevail.

To view matters quantitatively, one looks again at Equations (3.57) and (3.58):

$$[dU]_n = C_V dT = -pdV + TdS \tag{3.65}$$

The ideal gas law and kinetic theory enable two substitutions:

$$[dU]_n = C_V dT = \frac{3}{2}nR \cdot dT = -\frac{nRT}{V} \cdot dV + TdS \tag{3.66}$$

Since the 300 K conditions maintain, U remains constant. Therefore,

$$dU = 0 = \frac{-nRT}{V} \cdot dV + TdS \tag{3.67}$$

Rearrangement and T cancellation lead to:

$$[dS]_{T,n} = \frac{nR}{V} \cdot dV \tag{3.68}$$

whereupon

$$S_j - S_{initial} = \int_{V_{initial}}^{V_j} nR \frac{dV}{V} \tag{3.69}$$

$$= nR \cdot \log_e \left(\frac{V_j}{V_{initial}} \right)$$

Equation (3.69) offers a handle on the entropy change for each gas compartment as the wall moves. The right-side gas expands, while the left is compressed. Therefore, the entropy changes are positive and negative for the right and left, respectively. Again, the entropy changes for a system need not be positive everywhere.

The lower panel of Figure 3.8 shows the total entropy change as a function of the pressure difference between the gas compartments. Important is that the change demonstrates a single maximum value when the pressure difference is zero. Thus, mechanical equilibrium applies when the pressure is uniform throughout the system. Attaining this brings the maximum entropy allowed by the circumstances. Note the mechanical equilibrium to be as robust as thermal. If the wall slides too far accidentally, then the entropy increase will not be the maximum possible. Not to worry. The pressure gradient so generated will enable work to be transferred from left to right and restore the entropy maximum along the way. The equilibrium state remains anything but singular. It indeed encompasses the states in the vicinity of the maximum entropy one.

One next considers Figure 3.9, which illustrates a third composite system. There are two features to note. First is that the interior wall is porous and permits the exchange of gas between left and right compartments. The second is the presence of two components: 1.00 mole of neon on the left initially and 1.00 mole of helium on the right. Let the temperature and pressure be uniform and each compartment be of volume 1.00 meter³.

Figure 3.9 illustrates a case of nonzero chemical gradients. Taking the temperature to be 300 K, the initial pressure of neon on the left is:

$$P_{initial}^{(Ne,L)} = \frac{n_{initial}^{(Ne,L)} RT}{V_L} \tag{3.70}$$

$$\approx 2.49 \times 10^3 \, \text{pascals}$$

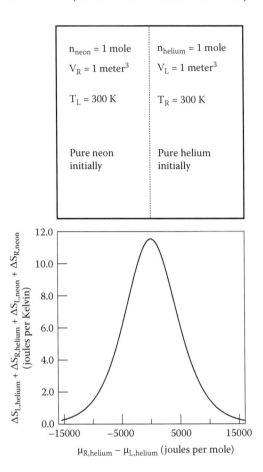

FIGURE 3.9 Composite systems and chemical equilibrium. The neon and helium mix until the chemical potential of each is uniform throughout. The lower frame shows the total entropy change as a function of the helium potential difference between the two sides. The change is maximum when the difference is zero. An equivalent plot can be constructed regarding the neon chemical potential.

This pressure is matched by the 300 K helium on the right side. Since the dividing wall is leaky, the gases will mix over time. This will not affect T or p on account of the gas ideality. Rather it will be the mole fractions and partial pressures that will change and indeed converge toward equilibrium values.

One focuses not on the internal energy because it remains fixed for both compartments. The key quantity is instead the free energy introduced by Equation (3.35). For a one-component system, differentiation of Equation (3.35) and combining with Equation (3.40) leads to:

$$dG = Vdp - SdT + \mu dn = \mu dn + nd\mu \tag{3.71}$$

and then

$$\frac{V}{n} \cdot dp - \frac{S}{n} \cdot dT = d\mu \qquad (3.72)$$

after canceling terms and moving n to the opposite (left) side. For an ideal gas at constant temperature, Equation (3.72) reduces to:

$$[d\mu]_T = \frac{RT}{p} \cdot dp \qquad (3.73)$$

whereupon

$$\mu_2 - \mu_1 = \int_{p_1}^{p_2} \frac{RT}{p} \cdot dp = f(T) + RT \log_e (p_2) - RT \log_e (p_1) + C_1$$

(3.74)

$$= f(T) + RT \log_e \left(\frac{p_2}{p_1} \right) + C_1$$

$f(T)$ is a function of temperature only while C_1 is an integration constant. It is customary to take C_1 to be zero and the lower integration limit p_1 as the unit pressure, for example $p_1 = 1.00$ pascal; and to apply different notation for $f(T)$. Thus the chemical potential for an ideal gas component is traditionally stated as

$$\mu_i = \mu_i(T,p) = f_i(T) + RT \log_e \left(\frac{p_i}{1.00} \right)$$

(3.75)

$$= \mu_i^o(T) + RT \log_e (p_i)$$

where the subscripts refer to the component identity, and the logarithm argument gives the appearance of having pressure units. The tacit agreement is that the unit pressure at a particular temperature (T) has been assigned as a reference state, in this case for the ith component. Note again that the first term of Equation (3.75) depends on temperature only; it is referred to as the standard chemical potential for component i. The term *standard* is denoted by the superscript in $\mu_i^o(T)$.

Now there are four intensive potentials relevant to Figure 3.9:

$$\mu_{Ne}^{left-side}(T,p) = \mu_{Ne}^o(T) + RT \log_e \left(p_{Ne}^{left-side} \right) \qquad (3.76A)$$

$$\mu_{Ne}^{right-side}(T,p) = \mu_{Ne}^o(T) + RT \log_e \left(p_{Ne}^{right-side} \right) \qquad (3.76B)$$

$$\mu_{He}^{left-side}(T,p) = \mu_{He}^o(T) + RT \log_e \left(p_{He}^{left-side} \right) \tag{3.76C}$$

$$\mu_{He}^{right-side}(T,p) = \mu_{He}^o(T) + RT \log_e \left(p_{He}^{right-side} \right) \tag{3.76D}$$

There are an equal number of entropy terms, each obtained from the relation

$$S = -\left(\frac{\partial G}{\partial T} \right)_{p,n} = -\left(\frac{\partial [n \cdot \mu]}{\partial T} \right)_{p,n} = -n \cdot \left(\frac{\partial \mu}{\partial T} \right)_{p,n} \tag{3.77}$$

Equation 3.77 applied to each potential yields:

$$S_{Ne}^{left-side} = -n_{Ne}^{left-side} \cdot \left[\left(\frac{d\mu_{Ne}^o(T)}{dT} \right) + R \cdot \log_e \left(p_{Ne}^{left-side} \right) \right] \tag{3.78A}$$

$$S_{Ne}^{right-side} = -n_{Ne}^{right-side} \cdot \left[\left(\frac{d\mu_{Ne}^o(T)}{dT} \right) + R \cdot \log_e \left(p_{Ne}^{right-side} \right) \right] \tag{3.78B}$$

$$S_{He}^{left-side} = -n_{He}^{left-side} \cdot \left[\left(\frac{d\mu_{He}^o(T)}{dT} \right) + R \cdot \log_e \left(p_{He}^{left-side} \right) \right] \tag{3.78C}$$

$$S_{He}^{right-side} = -n_{He}^{right-side} \cdot \left[\left(\frac{d\mu_{He}^o(T)}{dT} \right) + R \cdot \log_e \left(p_{He}^{right-side} \right) \right] \tag{3.78D}$$

It is worth mentioning that the neon initially contributes zero pressure to the right side. Its free energy and entropy contributions—extensive quantities—are thereby zero. However, the chemical potential—an intensive quantity—of right-side neon is not zero likewise, but rather negative infinity at the start. This is gathered from the limit properties of exponentially related variables, that is:

$$x = e^y: y \to -\infty \text{ as } x \to 0 \tag{3.79}$$

It follows that

$$\mu_i(T,p) = \mu_i^o(T) + RT \log_e(p_i) \to -\infty \tag{3.80}$$

in the limit of $p_i \to 0$. The equivalent statement holds for the initial amount of helium considered for the left-side compartment. The point is that the neon diffuses toward the right where its chemical potential is lower; the helium mixes by moving left where its potential is lower. The potential disparities are infinite at the start. They become finite as the mixing progresses and the entropy increases.

Equations (3.78A) to (3.78D) enable computation of the entropy changes for the helium and neon in each compartment. In the first two exercises, the entropy decreased on one side and increased on the other. Mixing is quite a different process in that the entropy changes are generally positive all around.

The lower portion of Figure 3.9 shows the total entropy as a function of the helium left- and right-potential difference. The behavior is mirrored by the neon. It is evident that the entropy is maximum when the chemical potential difference between the left and right sides for a component is zero. The condition of chemical equilibrium is obtained when the potential of each component becomes uniform throughout the system. The smoothing of μ gradients indeed steers all parties toward the maximum entropy state. Note that μ for one component need not equal μ of another. It is instead the spatial uniformity of a component's potential that is the signature of chemical equilibrium. Note further that the chemical equilibrium cannot be obtained unless the system is also in thermal and mechanical equilibrium; chemical equilibrium is subsidiary to the latter two. Let us not overlook critical and subtle features. The state of chemical equilibrium is not singular as it involves the states near the maximum entropy one. Further, when the conditions conform to the maximum entropy, a minimum number of variables are able to describe the system. Any and all strays from maximum entropy mandate additional information in the facts-and-data sense in order to portray the system. Helium and neon are the most inert of gases. Even so, the significance of this exercise will figure again in Chapter 7 concerning chemical reactions. The deviations from maximum entropy are addressed further in the next section.

3.3 SYSTEMS AND STATE POINT INFORMATION

Equations of state convert input variables to output. The latter anticipate properties of a system that can be checked by experiment. The infrastructure of Sections 3.1 and 3.2 is famously applicable to systems and input–output conversions. The proviso is that the conditions conform to equilibrium—the most probable states are manifest, there is zero available work, and so forth.

The foregoing statements are correct up to a point. They require additional elaboration because of the issues raised in Figure 3.6 and the equilibration exercises of the previous section. In particular, if ever a system attains the maximum entropy state, deviations remain possible and indeed transpire ever after. A maximum entropy system is not static, but rather is pushed and pulled by nature repeatedly. This is the case even if V, n, T, or other control variables are held fixed as best as possible by the chemist. By themselves, thermodynamic variables provide vital facts and data information; this was a point introduced in Chapter 1. Yet it is the pushing and pulling due to momentary gradients that confer information in the statistical sense. But then, how much information? This is addressed in a simple example.

Let the ideal gas equation be used to convert experimentally measured V, n, and T into output; let the result be 5774 pascals. Upon the fourth measurement, the chemist finds the pressure dial to read 5654 pascals—the ideal gas equation overestimated by 120 pascals. Then if p is measured subsequent times, the results should equate with the ideal number minus a correction of 120 pascals. If the van der Waals equation is used instead to anticipate p, a similar situation arises. The correction should be smaller as account has been taken of the nonideality.

The preceding is not 100% accurate because it implies that repeated measurements afford zero information. It implies that the probability of observing $p = 5654$

pascals is exactly 1, given fixed V, n, and T. There is more to the story due to the pushing and pulling about the maximum entropy state.

One considers the unusual system in Figure 3.10. A single neon atom has been represented in highly magnified form in a double-chamber container. The exterior wall is diathermal and enables the temperature to be held constant by a surrounding bath. According to kinetic theory (cf. Equation 3.58), the atom demonstrates average energy U of

$$
\begin{aligned}
U &= \left(\frac{3}{2}\right) nRT \\
&= \left(\frac{3}{2}\right) \cdot \left(\frac{1}{6.023 \times 10^{23}}\right) \cdot RT = \left(\frac{3}{2}\right) \cdot \left(\frac{R}{6.023 \times 10^{23}}\right) \cdot T \\
&= \left(\frac{3}{2}\right) \cdot k_B T
\end{aligned}
\tag{3.81}
$$

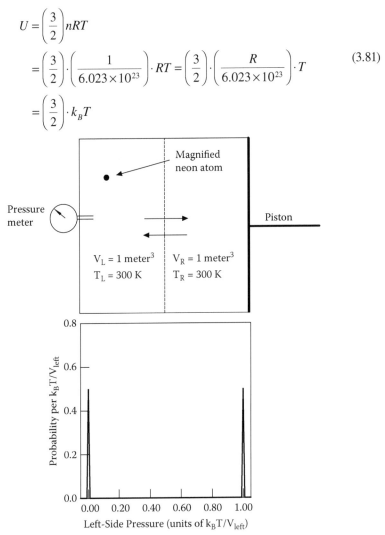

FIGURE 3.10 Composite systems and thermodynamic uncertainty. The meter attached to the left compartment registers a signal, depending on whether the neon atom is present or absent. The atom is free to drift between the compartments. The lower panel shows an idealized probability density function. The effects of moving the piston right to left are considered in Exercise 3.12 at the end of the chapter.

Yet as in Figure 3.9, a porous wall enables the atom to drift between left and right. If the compartments are of equal volume, the atom is just as likely to reside in one as the other. A barometer attached to the left chamber will register the pressure of either zero or something in the neighborhood of

$$p = \frac{1 \cdot k_B T}{V_{left}} \qquad (3.82)$$

Figure 3.10 describes an idealized scenario of two possible states and where an elementary measurement is preceded by uncertainty. An experiment, by definition, aims at reducing uncertainty. Given the idealized probability density function (lower panel) allied with the pressure, a single measurement by the chemist avails 1.00 bit of Shannon-type information. The measurement addresses the question: Is the neon atom in the left-side compartment? The information is less if the compartments differ in any respect: volume, wall stickiness, and so forth.

The case of multiple gas particles should then be addressed. Represented in Figure 3.11 are systems that host $N \gg 1$ neon atoms in double-chamber containers at fixed temperature T. As in Figure 3.10, thermal energy turns the placement of each atom into a random variable. The probability of any particular atom residing at the left or right is proportional to the volume set by the interior wall position. Two of infinite possible situations are illustrated in Figure 3.11. For system 1, the probability of a neon atom being in the right compartment is three times that of the left. For system 2, the left and right probabilities are equal.

Probability and pressure are usually denoted by the same symbol. To avoid confusion, the former will be represented by *prob* as in Chapter 2. Then for every atom in Figure 3.11, the following statements hold true:

$$prob(left) = \frac{V_{left}}{V_{left} + V_{right}} = \frac{V_{left}}{V_{total}} \qquad (3.83A)$$

$$prob(right) = \frac{V_{right}}{V_{left} + V_{right}} = \frac{V_{right}}{V_{total}} \qquad (3.83B)$$

If $N = 1$, there are two configurations possible: *left* (L) and *right* (R). For $N = 2$, there are four configurations allowed: LL, LR, RL, and RR. The atoms are identical, ignoring isotope details. Thus, subscripts tracking individual atoms are not appropriate—it is incorrect to write L_1R_2, L_2R_1, and so on. Accordingly for $N = 3$, there are eight possible configurations: LLL, LLR, LRL, LRR, RLL, RLR, RRL, and RRR, which can be written slightly more compactly as L^3, L^2R, L^2R, LR^2, RL^2, R^2L, R^2L, and R^3. While tedious, it is straightforward to identify the configurations for higher N. The feature to notice is the large number of equivalent ones, especially as N becomes large. The number of equivalent—and thus indistinguishable—configurations can be very large indeed.

A measurement of the left-side pressure will hinge on the gas that can access the barometer. To be sure, the maximum entropy state is where the left-side pressure

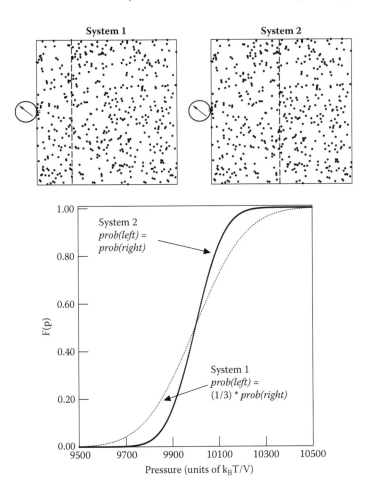

FIGURE 3.11 Composite systems and thermodynamic uncertainty for multiparticle systems. For system 1, the probability of an atom being in the right compartment is triple that of the left. For system 2, the left- and right-side probabilities are equal. The lower panel shows the probability distribution function associated with left-side pressure measurements.

exactly matches the right; ditto for the chemical potentials. The holes in the middle wall, however, allow deviations. There are many possible states in which p, μ for one compartment fail to match the intensive values for the other. The inequities never persist as they occasion forces (via gradients) that push the system back in the direction of maximum entropy.

There are multiple configurations that correspond to a given pressure in the left compartment. Unsurprisingly, some configurations are more plentiful and thus more likely to manifest than others. It is the binomial distribution that quantifies the probability of observing a configuration having x atoms in the left compartment and $N - x$ in the right:

$$prob(x, N - x) = \frac{N!}{x!\,(N - x)!} \cdot prob(left)^x \cdot prob(right)^{N-x}$$

$$= \frac{N!}{x!\,(N - x)!} \cdot prob(left)^x \cdot (1 - prob(left))^{N-x} \tag{3.84}$$

The distribution is akin to that encountered in the third peptide thinking exercise of Chapter 2—where the cell showed no selectivity in the placement of V and R. For the situation at hand, let the neon behave as an ideal gas. Equation (3.84) then quantifies the probability of recording a left-side pressure p of

$$p = \frac{x k_B T}{V_{left}} \tag{3.85}$$

Many values are possible: $\frac{0.21 \cdot N \cdot k_B T}{V_{left}}$, $\frac{0.28 \cdot N \cdot k_B T}{V_{left}}$, and so forth. The likelihood of each depends on N and the compartment volumes relative to one another.

Calculations assisted by the Stirling approximation (cf. Equations 2.48 and 2.49) illustrate the important points. The lower panel of Figure 3.11 shows the probability distribution for left-side pressure measurements. N has been arbitrarily set at 10^4 and $\frac{k_B T}{V_{total}}$ used as the unit measure; $V_{total} = V_{left}/prob(left)$ by Equation (3.83A). One observes that the distribution is narrower when the left and right compartments are of equal volume; contrarily the width is increased when $V_{left} < V_{right}$, that is, $prob(left) < prob(right)$. That the widths are nonzero at all means that uncertainty will precede measurements endeavored by the chemist. As in Figure 3.10, a thermodynamic measurement attaches to Shannon information in a nontrivial way. Figure 3.12 then shows the contrasts in the Figure 3.11 examples. If the chemist queries the left-side pressure at a resolution of $50 k_B T/V_{total}$, there are 20 or so states that will manifest frequently, and which he or she can discriminate. The plot shows the sum of weighted surprisals as a function of state index i—the lower p values correspond to lower i. There is about 30% more information, approximately 3.8 bits, trapped via a system 1 measurement compared with system 2. Clearly, when an equation of state is used to anticipate a quantity such as pressure, there are more issues at play than a correction term to add or subtract. To be precise, an equation of state furnishes an estimate of an average of a physical quantity subject to fluctuations.

There are additional quantitative details. The first moment of the Equation (3.84) distribution can be shown to equal $N \cdot prob(left)$. Hence, repeated measurements of the left-side pressure will lead to an average:

$$\langle p_{left} \rangle = N \cdot \frac{V_{left}}{V_{left} + V_{right}} \cdot \frac{k_B T}{V_{left}}$$

$$= N \cdot \frac{k_B T}{V_{total}} \tag{3.86}$$

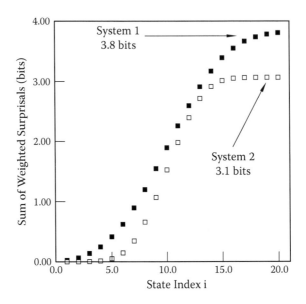

FIGURE 3.12 Weighted surprisal summation for left-side pressure states. The pressure measurements directed at system 1 offer greater information compared with system 2 because of the greater impact of fluctuations.

This is an important result. It shows that the left-side mean pressure *equals* that of the maximum entropy state for the system as a whole. Yet because the neon is able to diffuse freely between compartments, the left-side pressure—the right side, too—fluctuates interminably about the average. Any rise on the left stems from a drop on the right. This means that the value of every intensive property—μ, p, β_T, and so forth—is constant for neither compartment, although the same averages will be demonstrated by the two sides.

It is the width of the distribution that provides another critical element. The variance σ^2 of the distribution in Equation (3.84) can be shown to be

$$\sigma^2 = N \cdot prob(left) \cdot (1 - prob(left)) \qquad (3.87)$$

where the square root σ is the standard deviation (cf. Equation 2.55). In order to weigh the significance of fluctuations, one examines the ratio λ between σ and the average:

$$\lambda = \frac{\sqrt{N \cdot prob(left) \cdot (1 - prob(left))}}{N \cdot prob(left)}$$

$$= \frac{1}{\sqrt{N}} \sqrt{\frac{(1 - prob(left))}{prob(left)}} \qquad (3.88)$$

Figure 3.13 thereby presents a plot of $\lambda \times \sqrt{N}$ as a function of *prob(left)*. Recall that *prob(left)* depends on the volume disparities of the compartments. The plot reflects that

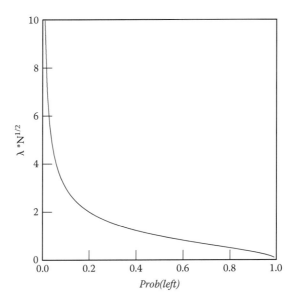

FIGURE 3.13 $\lambda \times \sqrt{N}$ versus *prob(left)*. The plot offers a reference for weighing the impact of fluctuations in a system containing N total number of particles. λ measures the ratio between the standard deviation and the average, while *prob(left)* equates with the fraction of the system volume.

if *prob(left)* ≈ 0.25 as for system 1, then $\lambda \times \sqrt{N} \approx 1.75$. Similarly, if *prob(left)* ≈ 0.50, then $\lambda \times \sqrt{N} \approx 1.00$ as in system 2. The plot serves as a reference for assessing all volume combinations. Along the way, it renders three important points. First is that λ changes nonlinearly with the volume disparity: $\lambda \approx 1.75/\sqrt{N}$, $1.00/\sqrt{N}$ for systems 1 and 2, respectively. Second is that the fluctuations wield greater impact the smaller the volume. How do small systems differ from large ones besides the obvious? The answers include the very disparate impact of fluctuations—their role generally looms large in small systems and outsized in very small (e.g., nanometer scale) systems. Third, the impact of fluctuations diminishes with increasing N. This tells us that thermodynamic measurements yield nonzero Shannon-type information—there is uncertainty to remove because a state point is not infinitely sharp. However, the amount falls as the number of atoms or molecules that compose the system is increased.

Emphasis should be added to the last few sentences. Figures 3.11 and 3.12 addressed the case of $N = 10^4$. This is roughly the number of molecules in a cubic sample of room air of volume (74 nanometers)^3. Let us consider a more reasonable size sample, say, 10^{-3} meter^3 (i.e., 1.00 liter) hosting Avogadro's number of molecules at room temperature. Taking the gas as ideal, the average pressure in any sector of the container will be:

$$p = \frac{N_{Av} \cdot k_B T}{V} = \frac{(6.02 \times 10^{23})(1.38 \times 10^{-23} \text{ joules}/K)(294\ K)}{0.00100 \text{ meter}^3} \approx 2.44 \times 10^6 \text{ pascals}$$

$$(3.89)$$

If a sector of interest comprises, say, one-third of the total volume, a quick look at Figure 3.13 informs one that

$$\lambda = \frac{1}{\sqrt{N}} \cdot \left[\frac{(1-prob)}{prob} \right]^{1/2}$$

(3.90)

$$= \frac{1}{\sqrt{6.02 \times 10^{23}}} \left[\frac{1-\left(\frac{1}{3}\right)}{\left(\frac{1}{3}\right)} \right]^{1/2} \approx 1.82 \times 10^{-12}$$

In turn,

$$\sigma_p = \lambda \cdot \langle p \rangle \approx (1.82 \times 10^{-12})(2.44 \times 10^6 \text{ pascals}) \approx 4.44 \times 10^{-6} \text{ pascals} \quad (3.91)$$

The pressure fluctuations of the sector are miniscule! In order to register them, the chemist requires a barometer that is accurate to about one part in 10^{12}—the barometer would have to register p to 12 significant figures or better. Deviations from average that exceed $3\sigma_p$ occur in less than 0.5% of a sample population. If the chemist had such high-precision instrumentation available, it would offer at most:

$$I = \frac{-1}{\log_e (2)} [0.995 \cdot \log_e (0.995) + 0.005 \cdot \log_e (0.005)] \approx 0.0454 \text{ bits} \quad (3.92)$$

per measurement. The values of thermodynamic state variables are vital facts and data information. Their measurement typically offers only sparse information in the statistical sense.

The major points of Chapter 3 are:

1. Thermodynamics is supported by an infrastructure of multivariable functions and equations of state. These apply the tools of integral and differential calculus. Equations of state are restricted, however, to the very special conditions of equilibrium.
2. Multivariable functions enable the state points of a system to be located in different coordinate systems: pV, pT, VT, and so forth. Their application, however, is not equivalent to that practiced in pure calculus and analytic geometry. This is because fluctuations confer a certain wobble on the point position; they enable one state to convert freely to others, depending on the system size and composition. The equilibrium conditions are robust and restorative, and by no means static.
3. Fluctuations impose uncertainty, which, in turn, confers nonzero information in thermodynamic measurements. The amount is miniscule typically. The exceptions arise for systems of small V and sparse N.

3.4 SOURCES AND FURTHER READING

The infrastructure of thermodynamics has been thoroughly charted in texts too numerous to mention. The author's list of preferred books include ones by Kauzmann [4], Desloge [5], Klotz [6], Fermi [7], Landsberg [8], Kirkwood and Oppenheim [9], Lewis and Randall [10], Pitzer and Brewer [11], Callen [12], Zemansky [13], and Spanner [14]. Books by Stanley [15], Goodstein [16], and Hecht [17] do not focus on classical thermodynamics per se. They include, however, excellent single-chapter encapsulations of the subject. It was via Stanley's text that the author was first introduced to Legendre transforms and thermodynamic squares. These two subjects were reinforced by Desloge's book, and by presentations such as by Goldstein of classical mechanics [18]. Also to be noted is the succinct presentation of classical thermodynamics by Dunning-Davies [19].

This chapter appealed to a few results from kinetic theory. The book by Hecht lays thorough groundwork on this subject [17]. Regarding empirical equations of state, the author has found chemical engineering texts most instructive regarding history, applications, and limitations. Especially illuminating are the books by Kyle [2] and Jones and Dugan [3].

This chapter has discussed fluctuations at an elementary level. The role of fluctuations is presented at an advanced level in several places. Highly recommended are the works by Kittel and Kroemer [20], Landau and Lifshitz [21], Lavenda [22], and Berne and Pecora [23].

Calculus-based approaches to thermodynamics are not the only ones. Thermodynamic states admit descriptions using the tools of differential geometry. The reader is encouraged to consult the truly seminal (and challenging) work of Gibbs [24], and the years-later contributions of Tisza [25], followed by Weinhold [26]. It is not surprising that the mathematical structure of thermodynamics has been the subject of several treatises, such as by Giles [27]. Taking an unusual approach, Peusner has presented a large body of thermodynamics using the tools of network analysis [28].

Last, if information inspires much discussion, so does entropy. Recommended are the books by Dugdale [29], Serrin [30], and Denbigh and Denbigh [31]. The text by Craig approaches chemical thermodynamics primarily through the entropy state function [32].

3.5 SUGGESTED EXERCISES

3.1 Dimensions and units are important to all fields. (a) Show that (U/p) and (G/V) have dimensions of volume and pressure, respectively. (b) Show that (a/Rb) and (a/b^2) have dimensions of temperature and pressure, respectively, where a and b are the van der Waals constants, and R is the gas constant.

3.2 Refer to Table 3.3 listing select van der Waals constants. (a) Derive expressions that enable a and b to be converted to liter^2-atmosphere/mole^2 and liters/mole, respectively. (b) Does b/N_{Av} equate with the volume occupied by a single gas molecule? Please discuss.

3.3 Figure 3.9 pertained to 1.00 mole each of helium and neon. The exercise focused on the mixing of gases and maximization of entropy. (a) Show that the maximum total change in the entropy for the composite system is $\Delta S_{max} = 2$ moles $\times R \times \log_e(2)$. (b) The chemical potential for an ideal gas includes a term that depends only on temperature. In arriving at ΔS_{max}, what is the fate of the $\mu^o(T)$ terms?

3.4 Revisit Chapter 2. Derive Equation (2.18) regarding the entropy of mixing for ideal gases.

3.5 The binomial distribution was presented in Equation (3.84). (a) Show that the first moment—the first cumulant—is $N \cdot prob(left)$. (b) Derive Equation (3.87) regarding the second cumulant.

3.6 For a $\kappa = 1$ system, H depends explicitly on p, S, and n. (a) Is H concave upward or downward with respect to p? Let the same question apply to S. (b) Establish the Maxwell identities based on $H(p, S, n)$:

$$\left(\frac{\partial V}{\partial S}\right)_{p,n} = \left(\frac{\partial T}{\partial p}\right)_{S,n} \quad \left(\frac{\partial T}{\partial n}\right)_{p,S} = \left(\frac{\partial \mu}{\partial S}\right)_{p,n} \quad \left(\frac{\partial V}{\partial n}\right)_{S,p} = \left(\frac{\partial \mu}{\partial p}\right)_{S,n}$$

3.7 Examine the Legendre transform of $H(p, S, n)$ with respect to p, S, and n. For each case, establish the simplest form analogous to Equations (3.33) and (3.34).

3.8 (a) Use the van der Waals equation to obtain a form for the isothermal compressibility, Equation (3.25). Do likewise for the Dieterici equation. (b) Refer to the van der Waals constants for argon. Compute and graph van der Waals β_T as a function of temperature: Let $n = 2.00$ moles, $V = 0.00100$ meter^3, and 200 K $\leq T \leq$ 500 K. What portion of the graph is best approximated by ideal gas β_T?

3.9 Use the van der Waals equation to obtain a form for the thermal expansivity, Equation (3.24). Do likewise for the Dieterici equation. (b) Refer to the van der Waals constants for xenon and plot α_p as a function of temperature. Let $n = 2.00$ moles, $V = 0.00100$ meter^3, and 200 K $\leq T \leq$ 500 K. What part of the graph is best approximated by ideal gas α_p?

3.10 For a $\kappa = 1$ system, compute:

$$U - \left(\frac{\partial U}{\partial V}\right)_{S,n} \cdot V \quad - \left(\frac{\partial U}{\partial S}\right)_{V,n} \cdot S \quad - \left(\frac{\partial U}{\partial n}\right)_{V,S} \cdot n$$

3.11 Consider the apparatus of Figure 3.7; only let the right compartment contain 1.00 mole of nitrogen gas ($N_2(g)$) initially at temperature 400 K. The left side contains 1.00 mole of neon at initial temperature 300 K. Construct a plot that shows the total change in the entropy as a function of the temperature difference between the left and right compartments. Take the heat capacity of the nitrogen to equate with:

$$C_V = \left(\frac{5}{2}\right) \cdot 1.00 \text{ mole} \times R$$

The extra capacity, compared with neon, derives from the end-over-end rotations of the molecules.

3.12 The flip of a coin leads to 1.00 bit of information trapped at the expense of work and dissipated heat. Confining the neon atom of Figure 3.10 to the left compartment via an isothermal compression also purchases 1.00 bit. (a) Perform a thermomechanical experiment at home: Flip a quarter and estimate the work transferred to the translational and rotational degrees of freedom. Do likewise with a dime. (b) Compare the work with that required for isothermal compression of the Figure 3.10 system, that is, moving the piston right to left and stopping at the center wall. Which system—the quarter, dime, or neon—affords the most expensive information? The least? Discuss the significance of these results. Note in particular how the information is purchased only by supplying work, dissipating heat, and lowering the system entropy.

3.13 Consider again the apparatus of Figure 3.11. Let the system consist of 5000 nitrogen molecules plus 5000 neon atoms at temperature 300 K. (a) Construct and plot the probability distribution function for C_V of the left-side container. (b) Nitrogen and neon have different specific heats. Should the plot of part (a) be bimodal? Please discuss. (c) Let the chemist measure left-side C_V with sufficient resolution to detect 0.10σ deviations from average. How many bits of information are trapped?

3.14 Consider an ideal gas sample of $N = 10^4$ molecules in a 1.00 meter^3 container at 300 K. What size sectors correspond to $\lambda = 0.0500, 0.100$, and 1.00?

3.15 Consider 1.00 mole of argon in a 1.00×10^{-3} meter^3 container at 200 K. (a) What is the average pressure estimated by the van der Waals equation for a sector equal to one-tenth of the total gas volume? (b) What is σ_p estimated for the sector?

REFERENCES

[1] Weast, R. C., ed. 1972. *Handbook of Chemistry and Physics*, p. D146, Chemical Rubber Co., Cleveland, OH.

[2] Kyle, B. G. 1999. *Chemical and Process Thermodynamics*, Prentice Hall PTR, Upper Saddle River, NJ.

[3] Jones, J. B., Dugan, R. E. 1996. *Engineering Thermodynamics*, Prentice Hall, Englewood Cliffs, NJ.

[4] Kauzmann, W. 1967. *Thermodynamics and Statistics*, W. A. Benjamin, New York.

[5] Desloge, E. A. 1968. *Thermal Physics*, Holt, Rinehart, and Winston, New York.

[6] Klotz, I. 1964. *Introduction to Chemical Thermodynamics*, W. A. Benjamin, New York.

[7] Fermi, E. 1956. *Thermodynamics*, Dover, New York.

[8] Landsberg, P. T. 1978. *Thermodynamics and Statistical Mechanics*, Dover, New York.

[9] Kirkwood, J. G., Oppenheim, I. 1961. *Chemical Thermodynamics*, McGraw-Hill, New York.

[10] Lewis, G. N., Randall, M. 1923. *Thermodynamics and the Free Energy of Chemical Substances*, McGraw-Hill, New York.

[11] Pitzer, K. S., Brewer, L. 1961. *Thermodynamics*, McGraw-Hill, New York.

[12] Callen, H. B. 1960. *Thermodynamics: An Introduction to the Physical Theories of Equilibrium Thermostatics and Irreversible Thermodynamics*, Wiley, New York.

[13] Zemansky, M. W. 1968. *Heat and Thermodynamics: An Intermediate Textbook*, McGraw-Hill, New York.

[14] Spanner, D. C. 1964. *Introduction to Thermodynamics*, Academic Press, London.

[15] Stanley, H. E. 1971. *Introduction to Phase Transitions and Critical Phenomena*, Oxford University Press, New York.

[16] Goodstein, D. L. 1985. *States of Matter*, Dover, New York.

[17] Hecht, C. E. 1990. *Statistical Thermodynamics and Kinetic Theory*, W. H. Freeman, New York.

[18] Goldstein, H., Poole, C., Safko, J. 2002. *Classical Mechanics*, 3rd ed., chap. 8, Addison Wesley, San Francisco.

[19] Dunning-Davies, J. 1996. *Concise Thermodynamics: Principles and Applications in Physical Science and Engineering*, Albion, Chichester.

[20] Kittel, C., Kroemer, H. 1998. *Thermal Physics*, 2nd ed., W. H. Freeman, New York.

[21] Landau, L. D., Lifshitz, E. M. 1958. *Statistical Physics*, chap. 22, Pergamon Press, London.

[22] Lavenda, B. H. 1991. *Statistical Physics*, Wiley, New York.

[23] Berne, B. J., Pecora, R. 2000. *Dynamic Light Scattering with Applications to Chemistry, Biology, and Physics*, chap. 10, Dover, New York.

[24] Gibbs, J. W. 1948. *The Collected Works of J. Willard Gibbs; Part I: Thermodynamics*, Yale University Press, New Haven, CT. (The first three sections make for dense but illuminating reading for the thermodynamics enthusiast.)

[25] Tisza, L. 1966. *Generalized Thermodynamics*, MIT Press, Cambridge, MA.

[26] Weinhold, F. 1976. Geometric Representation of Equilibrium Thermodynamics, *Acc. Chem. Res.* 9, 232. See also, Weinhold, F. 1975. Metric Geometry of Equilibrium Thermodynamics, *J. Chem. Phys.* 63, 2479, and papers that follow.

[27] Giles, R. 1964. *Mathematical Foundations of Thermodynamics*, Pergamon, Oxford.

[28] Peusner, L. 1986. *Studies in Network Thermodynamics*, Elsevier, Amsterdam.

[29] Dugdale, J. S. 1996. *Entropy and Its Physical Meaning*, Taylor & Francis, London.

[30] Serrin, J., ed. 1986. *New Perspectives in Thermodynamics*, Springer-Verlag, Berlin.

[31] Denbigh, K. G., Denbigh, J. S. 1985. *Entropy in Relation to Incomplete Knowledge*, Cambridge University Press, Cambridge.

[32] Craig, N. C. 1992. *Entropy Analysis: An Introduction to Chemical Thermodynamics*, VCH Publishers, New York.

4 Thermodynamic Transformations and Information

Fluctuations connect the states of a system by accident. By contrast, when variables are fine-tuned and coordinated with heat and work exchanges, the states are linked by design. We consider a special type of design, one that defines a locus of nearest-neighbor state points. The sequence of points marks a pathway in the manner of a computational program.

4.1 EQUILIBRIUM STATES, PATHWAYS, AND MEASUREMENTS

There were three themes of the previous chapter. First was the infrastructure for describing systems via potentials, state variables, and differentials. Second was the use of empirical equations to model systems at equilibrium. Third, was that fluctuations impose a nonzero width on every state point. A point anticipated by the ideal gas, van der Waals, or other equations of state is not infinitely sharp as in analytic geometry. Rather, a system demonstrates a range of pressure, density, and other properties. The fluctuations are as integral to the thermodynamic behavior as the average values.

These themes enjoy a reprise in Figure 4.1. The upper portion depicts a gas of volume V at equilibrium p, T, and μ. The infrastructure is bridged to empirical equations of state by differentials such as

$$p = -\left(\frac{\partial U}{\partial V}\right)_{S,n} = -\left(\frac{\partial A}{\partial V}\right)_{T,n} \tag{4.1}$$

$$V = +\left(\frac{\partial G}{\partial p}\right)_{T,n} = +\left(\frac{\partial H}{\partial p}\right)_{S,n} \tag{4.2}$$

There are many more examples. Because of energy exchanges, container shape fluctuations, and so on, the state point in the lower part of Figure 4.1 demonstrates a nonzero width. In turn, measurements of p and V at resolution windows Δp and ΔV furnish the chemist with a range of values about the average. Under most circumstances, Δp and ΔV exceed the state point width (as in the figure). The fluctuations are ever active but exert little impact when the systems are modest to large in size.

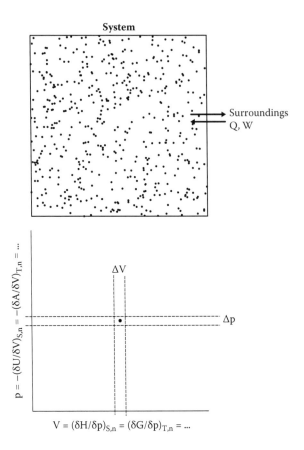

FIGURE 4.1 Reprise of Chapter 3 themes. The upper portion depicts a gas at equilibrium. Because of energy exchanges and other fluctuations, the state point for the system demonstrates a finite width as in the lower panel.

To introduce the next subject, another view of a state point must be considered. One imagines having a very high-resolution apparatus to record the changes in the Figure 4.1 system from one instant to the next. Let the point position be tracked by a hypothetical p,V microscope and stopwatch. This is only a thinking exercise, but the lesson is important and should be apparent. Properties such as p and V fluctuate about average values. Yet in doing so, the system does not express island points as in region A of Figure 4.2. Rather, the states (and their corresponding points in coordinate planes) are linked one neighbor to the next. The order of states accessed via the fluctuations offers a story as in region B. The pattern—more accurately, the lack of one—shows the state point width to arise from a jagged pathway. The distance and angle between each link depend on the relaxation of the system and coupling to the environment. These topics fall outside the scope of this book. What is emphasized, though, is that as the pathway is traced out, there occur zero changes in potentials U, H, A, and G on average; likewise for p, V, μ, and so forth. There occur momentary heat and work exchanges between the system and surroundings; these average to

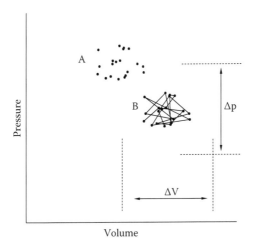

FIGURE 4.2 Pathway structure of a state point. A system does not demonstrate island points as in region A. Rather, the states are linked one to another by equilibrium fluctuations. A succession of points as in region B traces a pattern evocative of Brownian motion. The pathway structure is ordinarily undetectable given the typical measurement resolution window Δp and ΔV.

zero as well. The pathway structure is invisible to the chemist given everyday detection limits. As stated, this is strictly a thinking exercise.

Yet suppose the system was retrofitted in a way that enabled a special type of state change. The apparatus in Figure 4.3 offers one possibility. Let a piston allow mechanical work to be transferred between the system and surroundings. Let a heat bath and conducting walls enable the two-way flow of heat. The apparatus includes a thermometer and barometer for monitoring T and p. A position encoder on the piston keeps track of all volume changes. There is substantial equipment not shown. This includes clamp-on adiabatic walls that permit work-only exchanges. There is a heat bath ready and waiting for every temperature accessed in a transformation. The baths enable equilibrium to be maintained, both within the system and with the surroundings.

In short, the equipment enables fine-tuning of the thermodynamic state. And when the system is transformed reversibly, the state variables change in parallel, in conjunction with the energy exchanges. The system remains at equilibrium throughout the tuning process. It is traditional to represent transformations in a coordinate plane such as pV. As in Figure 4.4, a transformation marks the relocation of the state point from some initial to final position. The intervening coordinates mark a continuous and structured pathway. Note that irreversible operations do not accommodate the same luxury of graphing. These are complicated by gradients induced in one or more intensive properties. A nonequilibrium state really does not correspond to any location in the pV or other coordinate plane. A succession of states will not admit the curve in Figure 4.4 if even a single point fails to meet the equilibrium criteria. How is a reversible transformation different from an irreversible one? The answers include that the former is readily plotted; this cannot be said about the latter.

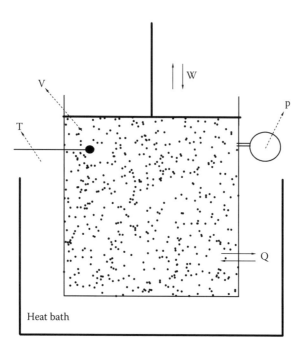

FIGURE 4.3 Tuning the thermodynamic state. A heat bath and piston enable energy exchanges and parallel tuning of the system variables.

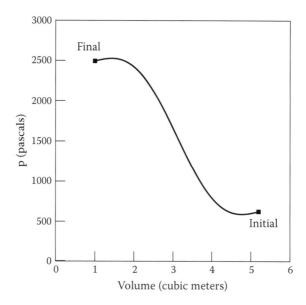

FIGURE 4.4 Reversible pathway in the pV plane. Each point of the path corresponds to an equilibrium state of the system.

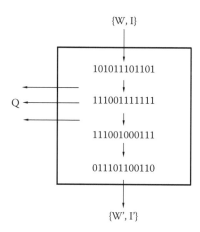

FIGURE 4.5 Computational devices and states. Work and heat exchanges plus information enable a device to transform a state such as 101011101101 to others along a programmed pathway. The exchanges are two-way between the system and surroundings.

Figures 4.3 and 4.4 illustrate scenarios with contemporary parallels. Figure 4.5 represents an electronic computer in highly schematic terms. Such a device also operates by work and heat exchanges. The transformations are coordinated as current and voltage states are converted along select pathways. As with all thermodynamic systems, computers do not exercise the transformations on their own. They require the information (I) of programs to supervise the state transitions and energy exchanges—heat (Q) and work (W). Computers are not equilibrium systems. Yet each of their operations is as deliberate as the pV ones of Figure 4.4. The information of a computer program is quantified in bits. It is important to consider the bits affiliated with pV and other thermodynamic transformations.

Computers operate along programmed pathways with information as the travel currency. Chapter 4 examines thermodynamic pathways for elementary systems in the same spotlight. They represent the most fundamental of system programs, long predating the ones driving laptops and smart phones. To set the stage, the next section describes the characteristics of reversible pathways.

4.2   A PRIMER ON REVERSIBLE TRANSFORMATIONS

Single-component ($\kappa = 1$) systems require three variables to locate the state point. At least one variable must be extensive such as V, n, S, and U. Leak-proof containers as in Figure 4.3 have fixed n. Their transformations are well portrayed on planes pV, TS, VT, and more. This is the message of Figure 4.6 where the curve of Figure 4.4 has been replotted as A in two and three dimensions (2D, 3D), taking n to be 1.00 mole. A companion B has been added to demonstrate how more than one path can connect an initial and final state. There are infinite possibilities as should be clear. All programmable devices, computers included, place no limit on the number of routes for taking one state to another.

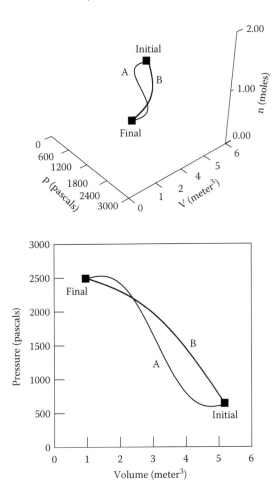

FIGURE 4.6 Reversible pathways in two and three dimensions. A corresponds to the pathway of Figure 4.4. B marks a different route with the same initial and final states.

The lower panel of Figure 4.6 shows how transformations are better discerned in 2D, compared with 3D. Not all situations afford this luxury. If the apparatus includes a needle valve or other mechanism for tuning n, it would fall upon 3D plots to tell the story.

A pathway combined with an equation of state provides alternative representations. For example, taking the Figure 4.6 system to be 1.00 mole of a monatomic ideal gas, the familiar equations apply:

$$p = \frac{nRT}{V} \tag{4.3}$$

$$U = \left(\frac{3}{2}\right) nRT = C_V T \tag{4.4}$$

Further, since the first law of thermodynamics mandates:

$$\left[dU\right]_n = -pdV + TdS \tag{4.5}$$

it follows that

$$dS = \frac{1}{T} \cdot dU + \frac{p}{T} \cdot dV$$

$$= C_V \cdot \frac{dT}{T} + nR \cdot \frac{dV}{V} \tag{4.6}$$

after combining the results of Equations (4.3) through (4.5). Integration of Equation (4.6) gives:

$$S_j - S_{initial} = C_V \cdot \log_e\left(\frac{T_j}{T_{initial}}\right) + nR \cdot \log_e\left(\frac{V_j}{V_{initial}}\right) \tag{4.7}$$

Applying Equations (4.4) and (4.7) to A and B leads to Figure 4.7. The upper panel portrays the system in the UV plane while the lower marks the TS structure. It is important that for each pV coordinate there corresponds a single point in the UV plane; likewise for TS. A one-to-one correspondence does not apply across the board, however. For example, an isothermal pathway in the pV plane for an ideal gas collapses to a single point in the UT plane. Different representations do not always afford equivalent knowledge.

p, V, T, S, and U are all functions of state. Hence, the net change of each is the same for A and B. This is not the case for W_{rec} and Q_{rec}, which are tied to the pathway structures. These quantities are obtained from the integrals

$$W_{rec} = -\int_{initial}^{final} p\,dV \tag{4.8}$$

$$Q_{rec} = +\int_{initial}^{final} T\,dS \tag{4.9}$$

W_{rec} equates with the area underneath the pV curves of Figure 4.6. Q_{rec} quantifies the area under the TS curves of Figure 4.7. Note that Q_{rec} follows alternatively from computing

$$Q_{rec} = \Delta U - W_{rec} \tag{4.10}$$

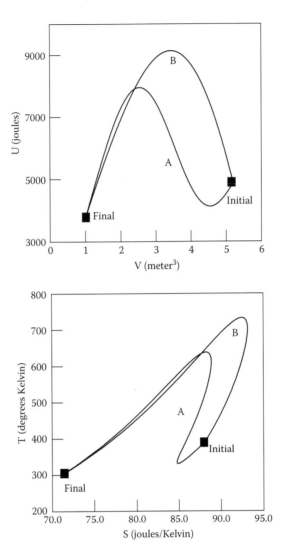

FIGURE 4.7   Reversible pathways in the *UV* and *TS* planes. The pathways correspond to A and B of Figure 4.6. The system has been taken to be 1.00 mole of monatomic ideal gas.

in accordance with the first law. Figure 4.8 illustrates the results of work and heat computations. Not surprisingly, B reflects greater work supplied by the surroundings to the system—there is greater pV area under B. At the same time, there is greater heat expelled ($Q_{rec} < 0$) as the system traverses B. W_{rec} and Q_{rec} are not functions of state in spite of their sum forming a state function.

There exist infinite programs for connecting state points. Some are special for their simplicity. Two are shown in Figure 4.9 by way of isochoric and isobaric pathways, also known as isochores and isobars. An isochore is notable because it entails

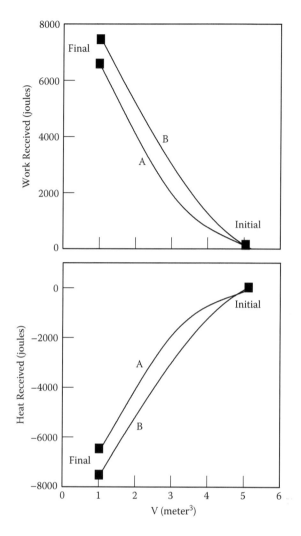

FIGURE 4.8 W_{rec} and Q_{rec} for reversible pathways. The pathways correspond to A and B of Figure 4.6. The system is 1.00 mole of monatomic ideal gas.

zero work; at constant V, $dV = 0$ and W_{rec} is zero by Equation (4.8). To direct a mechanical system to travel such a path requires locking the piston in place. Then only heat can be exchanged whereby p rises with addition to the system and falls with removal.

An isobaric pathway is no less special. In this case, heat is exchanged between the system and surroundings in either direction. Concomitant with the exchanges is the repositioning of the piston to maintain constant p. W_{rec} equates with a rectangle of area in the pV plane. Both W_{rec} and Q_{rec} scale linearly over the transformation.

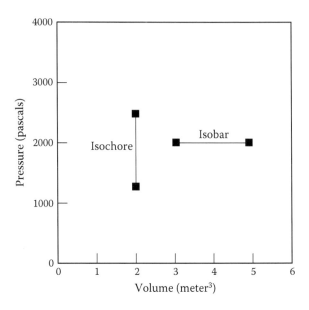

FIGURE 4.9   Special pathways: isochoric and isobaric.

In the pV plane, isochores and isobars appear as vertical and horizontal lines, respectively. This is not the case in Figure 4.10, which illustrates three isothermal paths in the upper panel. As always, the work and heat exchanges and variable tuning must be perfectly coordinated. If the piston inches downward, then $dW_{rec} > 0$. A compensating heat must be withdrawn (i.e., $dQ_{rec} < 0$) to preserve constant T. If the piston moves upward, heat injection is mandatory whereby $dQ_{rec} > 0$. The word *perfectly* is apropos. To preserve the equilibrium conditions, no new entropy must be created. The system must be describable by valid T, p, and μ at all stages.

The isotherm structure depends on the nature of the system. For an ideal gas:

$$p \cdot V = nRT \tag{4.11}$$

Thus $p \cdot V$ is a constant at all points. For isotherms described by the van der Waals equation:

$$\left(p + \frac{an^2}{V^2}\right) \cdot (V - nb) = nRT \tag{4.12}$$

Thus the invariant quantities are somewhat more complicated. As examples, the isotherms in Figure 4.10 apply to 1.00 mole each of ideal gas, xenon (Xe), and sulfur dioxide (SO_2). The data all pertain to the same temperature held at 200 K. The Xe and SO_2 pathways have been plotted via Equation (4.12) and the van der Waals constants of Table 3.3 of Chapter 3. It is apparent that the three examples become closely aligned at higher V. As would be expected, the excluded volume and attractive forces carry greater weight at lower V and higher p. This means that W_{rec} and Q_{rec} depend on

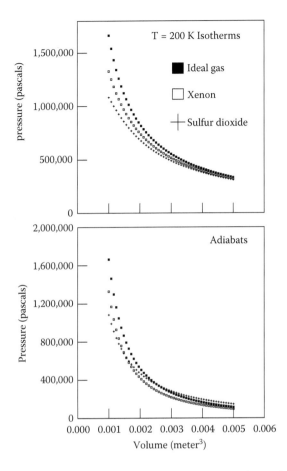

FIGURE 4.10 Special pathways: isothermal and adiabatic. The pathways for xenon and SO₂ have been computed using the van der Waals equation. Each pathway pertains to 1.00 mole of molecules.

the material in question. Of the three, SO₂ transfers the least work to the surroundings as its cohesive forces are strongest. For the ideal gas, it is simply that

$$W_{rec} = -Q_{rec} \qquad (4.13)$$

For van der Waals systems, W_{rec} and $-Q_{rec}$ are only approximately equal along isotherms.

Figure 4.10 includes constant entropy or adiabatic pathways in the pV plane. The lower panel transformations are more complicated as they call for extra attachments not represented in Figure 4.3. For a system to be directed along an adiabatic path, the heat reservoirs must be cast aside and the surrounding walls covered with insulation. As with isotherms, if the piston compresses the gas, $dW_{rec} > 0$. Since the "new energy" cannot leak out, the temperature rises sooner or later. If the piston moves upward, then $dW_{rec} < 0$. In this case, thermal energy is depleted and the temperature falls. Adiabatic transformations are unusual because they affect T, albeit with zero heat exchanged. The temperature changes can be drastic as in supersonic expansions and shock waves.

The mathematics of an adiabatic transformation is summarized as follows. For a closed ideal system unable to exchange heat, Equation (4.6) reduces to:

$$C_V dT + \frac{nRT}{V} \cdot dV = 0 \tag{4.14}$$

When T is repositioned, one obtains:

$$C_V \cdot \frac{dT}{T} + nR \cdot \frac{dV}{V} = 0 \tag{4.15}$$

Integration then arrives at a quantity that is invariant throughout the transformation, namely,

$$C_V \cdot \log_e(T) + nR \cdot \log_e(V) = \Lambda_1$$

$$= \log_e(T^{C_V}) + \log_e(V^{nR}) \tag{4.16}$$

$$= \log_e(T^{C_V} \cdot V^{nR})$$

It follows that

$$T^{C_V} \cdot V^{nR} = \exp[\Lambda_1] = \Lambda_2 \tag{4.17}$$

where Λ_1 and Λ_2 are constants dictated by the initial conditions. Equations (4.14) to (4.17) are restricted to ideal gases. Systems modeled by the van der Waals equation do not stray far. It can be shown that

$$U = C_V \cdot T - \frac{an^2}{V} \tag{4.18}$$

whereby

$$[dU]_n = C_V \cdot dT + \frac{an^2}{V^2} \cdot dV \tag{4.19}$$

Since

$$C_V = \left(\frac{\partial U}{\partial T} \right)_{V,n} \tag{4.20}$$

such a function has the same form for both ideal and van der Waals systems. Combining Equations (4.19) and (4.20) and the van der Waals equation leads to:

$$C_V \cdot \frac{dT}{T} + nR \cdot \frac{dV}{(V - nb)} = 0 \qquad (4.21)$$

The operations that led to Equation (4.17) can be applied to Equation (4.21) to yield:

$$T^{C_V} \cdot (V - nb)^{nR} = \exp[\Lambda_3] = \Lambda_4 \qquad (4.22)$$

Equation (4.22) can be viewed as an invariant signature of a van der Waals adiabat, which stays constant for the system in spite of the pV tuning.

The adiabats in Figure 4.10 apply to 1.00 mole each of an ideal gas, and for van der Waals Xe and SO_2. The initial T, p, and V have been taken to be the same as for the isotherms. The pathway differences are not trivial. The greatest work is delivered to the surroundings by the ideal gas; with expansion, there are neither attractive nor repulsive forces in play. The least work is transferred by the xenon; some of its thermal energy must be expended to overcome the attractive forces, while the monatomic character obtains a low heat capacity. The expansion work of SO_2 is less than that for the ideal gas because of attractive forces. There is compensation, however, as the triatomic character means a larger heat capacity. In particular, for SO_2:

$$C_V \approx \left(\frac{7}{2} \right) \cdot nR \qquad (4.23)$$

Note that the reductions in the SO_2 thermal energy content are not as sharp as for Xe. For the latter,

$$C_V \approx \left(\frac{3}{2} \right) \cdot nR \qquad (4.24)$$

To conclude this section, Figure 4.11 offers a sampling of cyclic pathways. These are rich in diversity and applications. In each case, the system's initial state is identical to the final. There are otherwise no limits placed on the intermediate states. For cyclic pathways, all quantities p, V, U, A, S, T, and so forth incur zero change ultimately, however convoluted the journey. In contrast to the examples of preceding figures, cyclic pathways do not correspond to single-valued functions. Their representation is established piecewise using two or more functions. Alternatively, they can be described by parametric equations in which the thermodynamic variables share

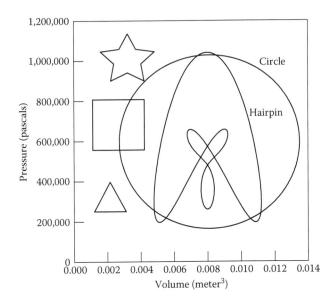

FIGURE 4.11 A sampling of cyclic pathways. The circle and hairpin are revisited in two subsequent figures.

a single-value dependence on η. For example, the hairpin pathway of Figure 4.11 conforms to:

$$V(\eta) = 10^{-3} \text{ meter}^3 \cdot \left[8 + 0.6\sin\left(\frac{2\pi\eta}{3}\right) + 1.6\sin\left(\frac{4\pi\eta}{3}\right) + 1.4\sin\left(\frac{6\pi\eta}{3}\right) \right] \quad \text{(4.25A)}$$

$$p(\eta) = 10^5 \text{ pascals} \cdot \left[5 + 0.9\cos\left(\frac{2\pi\eta}{3}\right) + 2.2\cos\left(\frac{6\pi\eta}{3}\right) \right.$$

$$\left. + 1.5\cos\left(\frac{8\pi\eta}{3}\right) + 0.8\cos\left(\frac{10\pi\eta}{3}\right) \right] \quad \text{(4.25B)}$$

where $0 \leq \eta \leq 3$.

Cyclic transformations impact daily life via heat engines and refrigerators. If a system is programmed for clockwise transit along any of the Figure 4.11 examples, heat from a surrounding reservoir is incrementally converted to work. The conversion efficiency is less than 100% by the second law of thermodynamics; there is always the need for a cooler reservoir to receive heat ejected (and thus wasted) by the system. Counterclockwise travel consumes work supplied by the surroundings. The results include heat withdrawal from one or more reservoirs and deposition in others. Refrigerators accomplish this night and day; the heat → work conversions of remote power plants enable them to operate. Life would be poorer without cyclic transformations and the programs that render them.

4.3 REVERSIBLE TRANSFORMATIONS AND INFORMATION

It is apparent that reversible pathways are an idealization. For a piston to transfer work, there must be a pressure gradient between the system and surroundings. Yet such a gradient must be infinitesimal for the equilibrium to maintain—it cannot be greater than ones generated by natural fluctuations. Likewise, for heat to flow spontaneously, there must be a temperature gradient. This also cannot exceed the gradients due to fluctuations. Reversible transformations accommodate thermal, mechanical, and chemical relaxation with every step. Given these stipulations, a reversible pathway requires as much as infinite time for travel—where the work and heat exchanges occur at zero power.

Idealizations provide lessons nonetheless. Reversible pathways are instructive because they offer ready approximations for real systems. Figure 4.8 presented Q_{rec} and W_{rec} along two pathways. The terminal values can be viewed as upper and lower limits, respectively, for real-life processes. Under normal circumstances, gradients other than via natural fluctuations create entropy. This "extra" entropy has to be expelled, and additional work is needed to land the system in the designated final state. The condition of reversibility provides a simple approach to complicated operations.

In a similar vein, reversible paths offer ready comparisons. For instance, the clockwise travel of each cycle in Figure 4.11 results in heat $\rightarrow$ work conversion. Which offers the greatest efficiency for 1.00 mole of gas? Taking each transformation as reversible and the gas as ideal would be a first step toward addressing the question.

Section 4.2 described a number of pathway fundamentals. Section 4.3 considers their information properties. A system does not initiate and trace a pathway on its own. It requires parallel operations specified by algorithms. Algorithms are rules and procedures for solving problems. They are wholly appropriate to thermodynamic venues. In each case, the problem is how can the system be transformed from a specified initial to final state along a demarcated path. The algorithm is imbedded in the sequence of state points. It enables one state to be converted faithfully to another without compromising the equilibrium. The algorithm calls for precisely executed variable tuning and work and heat transfers. In computation, algorithms are measured by their information content. How much information is expressed by thermodynamic pathways?

But there lie twin cruxes. A reversible path marks a locus of nearest-neighbor state points; there is virtually an infinite number in the general case. Information in Chapter 2 was grounded on finite state collections—coin faces, peptides, and so forth. By contrast, formulae such as Equations (4.25A) and (4.25B) specify infinitudes of pV states, and accordingly, fact and data information.

The second crux concerns probability. This was defined in Chapter 2 in terms of infinite trials and measurements. Once in place, the probability distributions never wavered for a system of interest.

Thermodynamic systems and pathways pose a very different situation. Assessing a point locus for information in the statistical sense requires the chemist to view the system in finite-resolution, objective terms. Information is quantified as a result of logical predictions of answers to yes or no questions. The basis for a prediction is one of reasonable belief on the chemist's part. This approach is subtly different from ones grounded on trials that are potentially infinite in number—coin tosses, peptide sampling, and the like.

Suppose that the chemist has superb knowledge of any of the pV pathways of this chapter. He or she is then well aware of the state boundaries of p and V along with their pairings. If, in addition, the chemist has knowledge of the equation of state, he or she can construct alternative representations such as UV and TS in Figure 4.7. With knowledge of the pathway, the chemist is cognizant of Q_{rec} and W_{rec} and all the changes in state functions. Pathway knowledge is a rich endowment of facts and data.

Yet by the condition of reversibility, the system must maintain equilibrium within itself and with the surroundings at all stages. Equilibrium states are like coins that lie flat on a table or like an isolated peptide to one critical extent: they offer zero testimony about the history or future. Therefore, while the chemist has a firm grasp of the collection of states, he or she is not spared the uncertainty that would precede any and all measurements. The chemist is pathwaywise; he or she is 100% sure that that measured p or V will lie *somewhere* within certain boundaries. But the chemist is ignorant of the state arrival, dwell, and departure times. There is doubt consequently attached to inquiries about the state at any instant.

Note how an individual state point and a pathway present contrasting scenarios. If the chemist knows average p and V as in Figure 4.1, but can extend measurements at resolution windows Δp and ΔV as illustrated, then all inquiries yield zero information in the statistical sense—p and V registered by the apparatus never alter. Matters are different for pathways such as in Figure 4.4 because the point relocations well exceed typical Δp and ΔV. If the chemist inquires about a system subject to such programming, and proceeds to address the question via a measurement, he or she reduces the uncertainty. He or she acquires information in the statistical sense at a cost of work dependent on the apparatus.

The examples of Figure 4.6 are plotted again in Figure 4.12. When the chemist probes p at resolution Δp, he or she is submitting questions fair and just such as

- Does the system pressure lie somewhere between 2400 and 2600 pascals?
- Does the pressure lie between 1500 and 1700 pascals?
- Does the pressure exceed 1800 pascals?
- Is the pressure less than 1000 pascals?

There are countless more. The chemist knows not to waste time with queries involving 10^8 or 10^{-8} pascals. The states corresponding to these values lie outside the domain prescribed by the algorithm. Prior to a measurement, the chemist lacks reason to hold one state more significant than another. He or she is also aware that reversible paths can be traveled in two directions—the term *reversible* is not incidental. Thus, the chemist lacks grounds for taking the program to be unidirectional in execution. Q_{rec} and W_{rec} plus the changes in U, S, p, and so forth are the same whether the transformation proceeds monotonically or in some back-and-forth fashion. All the chemist knows is that at some stage, the system, operating in tandem with the surroundings, departs from the initial state and eventually arrives at the final state. The program specifies which states

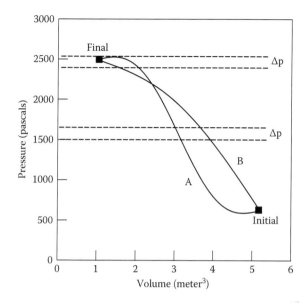

FIGURE 4.12   Pathways and queries. A and B correspond to Figure 4.6. When the chemist probes p at resolution window Δp, he or she is submitting questions such as: Is the system pressure between 2400 and 2600 pascals? Is p between 1500 and 1700 pascals?

are accessed. It is opaque, however, regarding the precise order and duration of accessing the states.

One hearkens to Figure 4.5. The states of a digital computer are dictated by a source program. The conversion of one state to another entails operations conducted in parallel—address queries, byte swaps, additions, and so on. Source programs enable a state to be visited and converted without limit; they are devoid of timing specifics, however. The upshot is that thermodynamic pathways are analogous to computational sequences to an important degree. Whether the chemist queries the state of a digital processor, or a gas programmed for heat and work exchanges, a subjective form of probability must be weighed to quantify the information. This form is a powerful tool in actuarial, financial, and disease-tracking research, among other fields. In thermodynamics, it is grounded on two essential postulates:

1. All states of a system along a pathway of interest are equally likely in occurrence.
2. The probability of observing, by objective measurement, a thermodynamic quantity X with value x somewhere in x and $x + \Delta x$ is proportional to the number of pathway states in that range.

Note the postulates pay no attention to the intricacies of measurement using barometers, thermometers, and so on. They further disregard the details of designing or executing a program. For example, the mechanics of trading heat reservoirs, or attaching or detaching adiabatic walls are immaterial.

The appeal to reasonable belief bridges information with pathway structure. For reversible A of Figure 4.12, the chemist regards the likelihood of observing p between 2400 and 2600 pascals to be greater than for 1500 to 1700 pascals. The rationale is that a more substantial fraction of the state points reside in the former neighborhood. Matters are different for the B program. Here comparable fractions manifest in the 1500 to 1700 and 2400 to 2600 pascal ranges. The chemist has no cause for thinking that one neighborhood is favored over the other, so he or she anticipates the system to express p states in the two ranges with about equal likelihood.

Probability admits mathematical descriptions. Thus, for every pathway and state variable of interest, there exist distribution functions denoted by $F_{X \leftrightarrow p}(p)$, $F_{X \leftrightarrow T}(T)$, $F_{X \leftrightarrow V}(V)$, and so forth. These are analogous to ones discussed in Chapter 2 for a random variable X having probability distribution $F_X(x)$. In thermodynamic venues, a function such as $F_{X \leftrightarrow p}(p)$ quantifies the reasonable belief likelihood of observing a system with pressure $\leq p$. Unlike everyday distributions—uniform, normal, and so on—$F_{X \leftrightarrow p}(p)$, $F_{X \leftrightarrow T}(T)$, and so forth are case specific and depend intricately on the pathway structure. There is really no limit to their diversity.

Information arrives via probability and surprisal values. It should be apparent how these will be obtained for reversible transformations. To assess the likelihood of observing p somewhere over a specified range, the chemist needs to tally the number of states programmed over that range. He or she will then divide the number by the total number of pathway states. The lessons of Chapter 2 will consequently apply; the greater the probability, the lower the surprisal and vice versa. The greater the uncertainty associated with a collection of states, the greater the information attached to a measurement. These statements hold for all thermodynamic quantities: V, T, p, μ, U, and so on.

There are additional issues. First, given a pathway of interest, the probabilities and surprisals do not follow at once. After all, the pathway structure depends on choices made regarding physical units and Cartesian axes. The circle and square of Figure 4.11 appear as an ellipse and rectangle if either axis, pressure or volume, is stretched. The same holds true if the transformations are plotted in, say, a torr–liter coordinate system. Second, probabilities and surprisals are not obtained, at least directly, from continuous functions and graphs. Summing the points specified by a function means evaluating a contour or line integral. A line integral for a p versus V function—lacking further treatment—has the unusual dimensions and International System of Units (SI units) of

$$\sqrt{p^2 + V^2} = \sqrt{\text{pascals}^2 + \text{meters}^6} \qquad (4.26)$$

Third, the pathway information established by the chemist is not singular but instead depends on the query nature and measurement window. Queries and measurements at resolution 1.00 pascal yield more information than at resolution 10^3 pascals. This also means that if a system was programmed for A or B of Figure 4.12, and yet probed at $\Delta p = 10^4$ pascals, then zero information would be obtained—all the states fall in the range 0 to 10^4 pascals and the barometer cannot discriminate one from another. The lesson is that information in the statistical sense must always be reported with the query and measurement details included. This is not a unique situation in thermodynamics as other properties call for fine-print attachments. As will be discussed in Chapter 7, an equilibrium constant must always be reported along with the temperature at which it was measured, the reaction stoichiometry, and the concentration or pressure units applied.

When quantifying pathway information, the procedure begins with the rescaling of the control variables. If the program specifies the states using p and V coordinates, then these govern all facets of the reversible pathway structure. They are the *control variables* whose rescaling is achieved as follows:

$$\bar{p} = \frac{p - p_{min}}{p_{max} - p_{min}} \qquad (4.27A)$$

$$\bar{V} = \frac{V - V_{min}}{V_{max} - V_{min}} \qquad (4.27B)$$

The subscripts label the values at the boundaries. Note that p_{min}, p_{max}, V_{min}, and V_{max} need not correspond to the initial and final states, although this is often the case. The rescaling leads to dimensionless quantities that span 0 to 1. The rescaling ensures that each control variable is weighted equally in placing the state points and defining the pathway structure. It is important that $\bar{p}, \bar{V}$ adopt the same values independent of the source units—pascals, meter³, torr, liters, and so forth. When applied to A and B of Figure 4.12, rescaling leads to the curves in the upper panel of Figure 4.13.

Information is a by-product of probability distributions. To construct, say, $F_{X \leftrightarrow \bar{p}}(\bar{p})$, one computes a contour length one of two ways. If $\bar{p}$ is a single-value function of $\bar{V}$, then dimensionless length Λ arrives by the integral

$$\Lambda = \int_0^1 d\bar{V} \cdot \sqrt{1 + \left(\frac{d\bar{p}}{d\bar{V}}\right)^2} \qquad (4.28)$$

Equation (4.28) would be appropriate to isothermal or adiabatic pathways, to name two.

If, instead, the pathway is grounded on parametric equations as in Equations (4.25A) and (4.25B), then the length is obtained via:

$$\Lambda = \int_{\eta_{initial}}^{\eta_{final}} d\eta \cdot \sqrt{\left(\frac{d\bar{V}}{d\eta}\right)^2 + \left(\frac{d\bar{p}}{d\eta}\right)^2} \qquad (4.29)$$

Equation (4.29) would apply to cyclic pathways and others where more than one p is paired with a given V. Equations (4.28) and (4.29) reflect well-known formulae from calculus. Note that while the reduced variables span 0 to 1, the length Λ can well exceed 1. As with computer programs, different pathways offer an assortment of lengths.

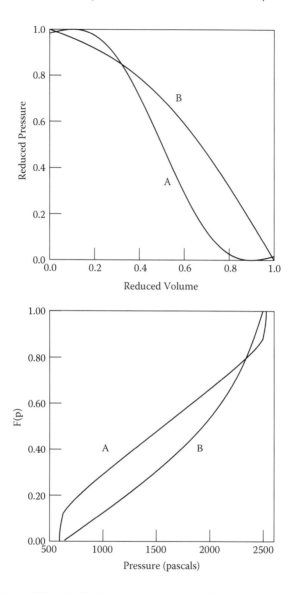

FIGURE 4.13 Probability distribution functions for reversible pathways. Shown are $F_{X \leftrightarrow p}(p)$ for A and B of Figure 4.6 and the previous figure.

The probability distributions are obtained as ratios of integrals. In each case, the denominator is Λ while the numerator is the *partial* contour length. Where Equation (4.28) applies, one computes:

$$F_{X \leftrightarrow p}(\bar{p}' \leq \bar{p}) = \frac{\left[\int_0^1 d\bar{V} \cdot \sqrt{1 + \left(\frac{d\bar{p}'}{d\bar{V}} \right)^2} \right]_{\bar{p}' \leq \bar{p}}}{\Lambda} \tag{4.30}$$

When Equation (4.29) applies, one computes:

$$F_{X\leftrightarrow\bar{p}}(\bar{p}'\leq\bar{p})=\frac{\left[\int_{\eta_{initial}}^{\eta_{final}}d\eta\cdot\sqrt{\left(\frac{d\bar{V}}{d\eta}\right)^2+\left(\frac{d\bar{p}'}{d\eta}\right)^2}\right]_{\bar{p}'\leq\bar{p}}}{\Lambda} \tag{4.31}$$

For both cases, the numerator is restricted to states $\bar{p}'\leq\bar{p}$. The calculations are nontrivial because several fragments of a pathway can contribute to the numerator. In all cases, the minimum value of the distribution function is 0 while the maximum is 1. Plots of $F_{X\leftrightarrow\bar{p}}(\bar{p})$ demonstrate a left to right increase that is governed entirely by the pathway structure. It is straightforward to convert the dimensionless variables back to physical ones for evaluation purposes; the lower panel of Figure 4.13 thereby illustrates $F_{X\leftrightarrow p}(p)$ specific to A and B. The distributions address a bounty of questions that the chemist can pose, for example:

1. How do average (and median) p for A and B compare? The answer follows by noting p for which $F_{X\leftrightarrow p}(p) = 0.50$. $<p>$ is registered as 1559 and 1933 pascals, respectively, for A and B.
2. For A and B programs, what is the likelihood of observing system $p \leq 1500$ pascals? The answer depends on the pathway structure: $F_{X\leftrightarrow p}(p \leq 1500$ pascals$) = 0.48$ for A and 0.31 for B.
3. At what value p_0 is the likelihood of observing system $p \leq p_0$ the same for A and B? The answer is $p_0 \approx 2270$ pascals where the $F_{X\leftrightarrow p}$ curves intersect.

Distribution functions are the vehicles for quantifying information, the amount depending on the query and measurement resolution. As with the control variables, the resolution window must be rescaled into a dimensionless form:

$$\Delta\bar{p}=\frac{\Delta p}{p_{max}-p_{min}} \tag{4.32}$$

Further, two or more pathways can only be compared—fairly that is—at identical physical resolution, say, 20 pascals. This will translate into different values of $\Delta\bar{p}$, given the disparities of pathway lengths, minima, and maxima.

The reasonable-belief probability allied with a specified range of states follows from a ratio such as

$$prob(\bar{p}\leq\bar{p}'<\bar{p}+\Delta\bar{p})=\frac{\left[\int_{\eta_{initial}}^{\eta_{final}}d\eta\cdot\sqrt{\left(\frac{d\bar{V}}{d\eta}\right)^2+\left(\frac{d\bar{p}'}{d\eta}\right)^2}\right]_{\bar{p}\leq\bar{p}'<\bar{p}+\Delta\bar{p}}}{\Lambda} \tag{4.33}$$

where the numerator is restricted to states $\bar{p}'$: $\bar{p}\leq\bar{p}'<\bar{p}+\Delta\bar{p}$. It is cumbersome to refer to states in such terms. It is more convenient to index (label) them via integers:

State	Index
$0 \leq \bar{p}' < \Delta\bar{p}$	1
$\bar{p} \leq \bar{p}' < \bar{p} + \Delta\bar{p}$	2
$\bar{p} + \Delta\bar{p} \leq \bar{p}' < \bar{p} + 2\Delta\bar{p}$	3
.	.
.	.
.	.
$(j-1) \cdot \Delta\bar{p} \leq \bar{p}' < j \cdot \Delta\bar{p}.$	j

Surprisals are then the by-product of the probabilities of Equation (4.33), as are the weighted surprisals:

$$S_j = -\log_2(prob(j))$$

$$= \frac{-1}{\log_e(2)} \cdot \log_e(prob(j)) \tag{4.34}$$

and

$$prob(j) \cdot S_j = \frac{-prob(j)}{\log_e(2)} \cdot \log_e(prob(j)) \tag{4.35}$$

The sum of weighted surprisals equates with the Shannon information for the variable X in question, in this case pressure:

$$I_{X \leftrightarrow p} = \frac{-1}{\log_e(2)} \cdot \sum_j prob(j) \cdot \log_e(prob(j)) \tag{4.36}$$

The results of evaluating Equation (4.36) are shown in Figure 4.14 for three resolution conditions. One finds A to demonstrate a slightly greater range of pressure states; this means a greater number of terms to sum in Equation (4.36). B expresses greater $I_{X \leftrightarrow p}$ by a few percent, however. This is because the pressure states in B are distributed in a less biased way. There is consequently greater uncertainty on the chemist's part prior to a measurement directed at a system programmed for B travel. The bias in A is acute near the minimum and maximum p. These states happen to be in the same vicinity as the initial and terminal. A programming avails less uncertainty prior to a measurement—and accordingly less information upon completion.

Anticipating the probability attributes of A and B is fairly straightforward. One inspects carefully the range of states and biases that have been programmed. Qualitative assessments are not always forthcoming, however. Figure 4.11 included

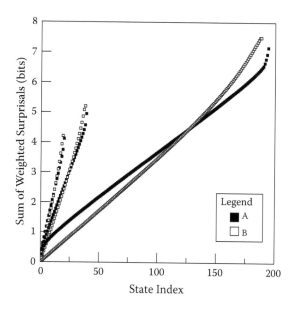

FIGURE 4.14 Information and pathways. Illustrated are the weighted surprisal sums for pressure states at three different resolutions. The pathways correspond to A and B of previous figures. At each query resolution, B expresses greater $I_{X \leftrightarrow p}$ on account of less bias in the state distribution.

hairpin and circle transformations. These have been rendered again in Figure 4.15 with a few accents added to make a point. In both cases, the same range of pressure states has been programmed. Which expresses greater $I_{X \leftrightarrow p}$? The answer is less than apparent. The circle reflects sizable state fractions in the accented neighborhoods of 2×10^5 and 10×10^5 pascals. The hairpin meanwhile concentrates on states near 4.2×10^5 pascals. The answer appears in the lower panel via the sum of weighted surprisals; both pathways have been evaluated at the same physical resolution. One finds $I_{X \leftrightarrow p}$ to be a dead heat—approximately 5.4 bits at resolution 1700 pascals. While the information is enhanced at higher resolution, the dead-heat status does not alter. The point is that thermodynamic programs do not have to be overly complex for the information properties to become nonobvious.

Contour integration is a major part of evaluating pathways on information grounds. Yet it should not surprise that the integrals can rarely be obtained by pencil-and-paper calculations. Numerical approximation is almost always necessary such as in the end-of-chapter exercises. This is not an obstacle with digital spreadsheets and some programming expertise. Sometimes the functions for a pathway are unavailable such as in Exercise 4. This is not a deal breaker. If a path is available by graph, it can be digitized readily. The rescaling steps can then be applied and the probability distributions follow quickly.

Significantly, thermodynamic transformations offer a rich variety of distributions. The constraints are that certain variables manifest only positive values: $V, p, T, N,$ $\alpha_p, \beta_T, \beta_S, C_V,$ and C_p. By the same token, because there exist special pathways—isothermal, isobaric, and so forth—as discussed in Section 4.2, there exist distributions

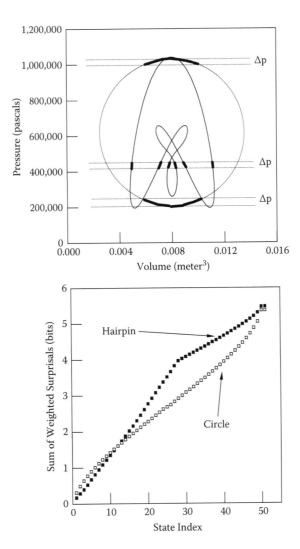

FIGURE 4.15 Cyclic pathways, state distributions, and information. The hairpin and circle of Figure 4.11 have been replotted with accents discussed in the text. The lower panel shows the weighted surprisal summation. $I_{X \leftrightarrow p}$ proves approximately equal for the two pathways.

that crop up frequently and are thus worthy of extra attention. These are discussed in the next section.

4.4 THE INFORMATION PROPERTIES OF REVERSIBLE PATHWAYS

A pathway is realized by work and heat exchanges, and the tuning of state variables. The actions are programmed in parallel and coordinated with the surroundings. Viewing pathway information as something separate is idealized. The perspective is valuable nonetheless in the way that computational programs are weighed in bit terms, quite apart from the transistor and diode circuits that materialize them. Information is always physical, as cited in Chapter 1. Yet it accommodates an analysis that places the mechanical details to one side, and focuses on the statistical structure of a state collection. It is the purpose of this section to summarize the properties of pathway information. Several have been apparent already.

The first is that individual states differ from pathways in that states offer almost zero information in the statistical sense. For an equilibrium system of appreciable size and particle number:

$$I_{X \leftrightarrow p} \approx I_{X \leftrightarrow V} \approx I_{X \leftrightarrow T} \approx I_{X \leftrightarrow \mu} \approx \cdots \approx 0 \tag{4.37}$$

This is on account of the minor effects of fluctuations. By contrast, pathways offer appreciable information in abundant variables.

The second property is that pathway information is independent of the travel direction. A contour length does not alter if the integration limits are interchanged. Thus $I_{X \leftrightarrow p}$ for A and B in Figure 4.12 does not alter if the initial and final states are switched. By the same token, the information of cyclic pathways (Figure 4.11) does not depend on whether the travel is clockwise or counterclockwise.

The third property is that different variables express different probability distributions and, in turn, information. For a pathway of interest, the chemist has a firm handle on several variables if he or she knows the equation of state. For example, taking the system of Figure 4.12 to be 1.00 mole of a monatomic ideal gas, the temperature is established at each point of A and B via:

$$T = \frac{p \cdot V}{1.00 \text{ mole} \cdot R} \tag{4.38}$$

Figure 4.16 illustrates $F_{X \leftrightarrow T}(T)$ for the A and B algorithms. The temperature bias is greater for A while the range is 25% larger for B. B thereby leads A in the information availed in temperature measurements by the chemist. Note the corollary: information expressed by control variables such as p and V begets additional information.

The fourth property lies in the contrasts between ideal and nonideal gases. The former demonstrate several properties beginning with the ideal gas law. An ideal gas further expresses C_V and C_p independently of temperature. For an ideal gas, the dependence of U and H is only on n and T and not at all on V. For an ideal gas, the response functions have elementary forms: $\alpha_p = T^{-1}$ and $\beta_T = p^{-1}$. A nonideal gas requires equations such as the van der Waals or Dieterici for description. In so doing, C_V, C_p, α_p, β_T, and other functions become quite a bit more complicated. In addition, potentials such as U and H automatically depend on V.

This fourth property accordingly marks the differences between ideal and non-ideal gases along information lines. An ideal gas offers zero $I_{X \leftrightarrow C_V}$ and $I_{X \leftrightarrow C_p}$ over all pathways, as long as n is constant. And for a gas to be ideal, zero $I_{X \leftrightarrow U}$, $I_{X \leftrightarrow H}$, and $I_{X \leftrightarrow \alpha_p}$ must apply to all isotherms. An ideal gas likewise expresses zero $I_{X \leftrightarrow \beta_T}$ for isobars. Nonideal gases lack these characteristics, and their pathway information is case specific as a result. Figure 4.10 presented isotherms for 1.00 mole each of

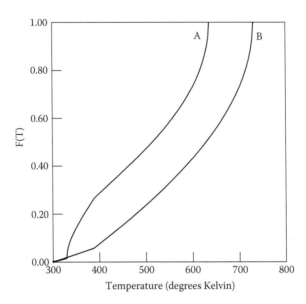

FIGURE 4.16 $F_{X \leftrightarrow T}(T)$ for A and B of previous figures. The temperature bias is greater for **A** while the range is approximately 25% larger for B. B presents the larger $I_{X \leftrightarrow T}$ under all resolution conditions.

monatomic ideal gas, xenon, and SO_2. The pathways for the latter two were established via the van der Waals equation and evaluated at $T = 200$ K. Initial and final V were identical for the three cases, while only the pressure domains differed.

The probability curves for the pressure states of these systems appear in Figure 4.17. They convey that if the chemist queried each system at identical physical resolution, the ideal gas would present the greatest $I_{X \leftrightarrow p}$. The Xe and SO_2 gas samples place second and third, respectively, regarding $I_{X \leftrightarrow p}$. Clearly, the information in pressure and other quantities depends on the component identity. The differences among gas samples are especially acute under low temperature, high density conditions.

The fifth property concerns the term *special*. Chapter 2 made the point that a handful of probability distributions are special because they apply so frequently in nature: uniform, exponential, and normal to name three. In a parallel way, special pathways are fixtures in thermodynamics as shown in Figures 4.9 and 4.10. The fifth property is that the special pathways for a system, ideal and otherwise, offer zero information in at least one state variable.

Isobars, isochores, isotherms, and isentropes express virtually zero $I_{X \leftrightarrow p}$, $I_{X \leftrightarrow V}$, $I_{X \leftrightarrow T}$, and $I_{X \leftrightarrow S}$, respectively. The qualifier virtually is added to acknowledge the effects of fluctuations—these always impose some uncertainty for a system. Isochores further express $I_{X \leftrightarrow W_{rec}} \approx 0$ while for isentropes, $I_{X \leftrightarrow Q_{rec}} \approx 0$. Not-so-special pathways such as A and B (Figure 4.12) offer information in all these quantities; they do not pose the *information economy* of special pathways. This does

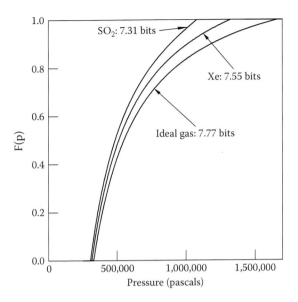

FIGURE 4.17 $F_{X \leftrightarrow p}(p)$ for the isotherms of Figure 4.10. The ideal gas avails the greatest $I_{X \leftrightarrow p}$. The resolution window Δp corresponds to 0.50% of the range of the SO_2 pathway.

not mean that special pathways are lacking in probability attributes. Quite the contrary: the sigmoidal and spike functions shown in Figure 4.18 apply to one or more state variables. The applicability of these functions is interesting. Both special and not-so-special pathways demonstrate state points in abundance. For the former, however, information in the statistical sense for one or more variables does not exceed that of a single point.

Special distributions carry extra weight in probability and statistics. Special pathways perform likewise in thermodynamics. By the fifth property, all the transformations of a closed system are special given that $I_{X \leftrightarrow n} = 0$.

The sixth property is that a pathway offers more than one flavor of information. The Kullback information (KI) in Chapter 2 followed from comparing two probability distributions. In thermodynamic venues, the KI_X quantifies the divergence of two programs for variable X, one serving as a reference for the other. For an example, the upper panel of Figure 4.19 illustrates two pV pathways sharing initial and final states. Over any region of pressure states, the fraction of 1 is not matched by 2. The statistical structures of the two programs are accordingly different. If the chemist anticipated the pressure states threaded by 1 according to 2, he or she would be in error to a degree measured by $KI_{X \leftrightarrow p}$. The calculation entails piecewise comparison of pathway length fractions. The numerator and denominator of each logarithm argument equate with probability values. $KI_{X \leftrightarrow p}$ arrives via the following weighted summation:

$$KI_{X \leftrightarrow p} = + \sum_{i=1}^{\Gamma} prob_1 \left(\bar{p}_i + \Delta \bar{p} \right) \cdot \log_2 \left[\frac{prob_1 (\bar{p}_i + \Delta \bar{p})}{prob_2 (\bar{p}_i + \Delta \bar{p})} \right] \qquad (4.39)$$

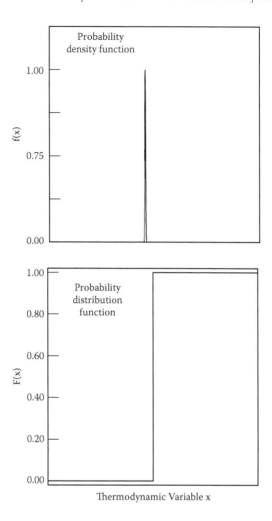

FIGURE 4.18 Probability functions for special pathways. The upper and lower panels show the probability density and distribution functions, respectively. A special pathway expresses zero information in one or more thermodynamic state variables.

where $\bar{p}_i + \Delta\bar{p}$ prescribe the widths and boundaries of each ith state. The number of terms Γ is set by the range and resolution:

$$\Gamma = \frac{p_{max} - p_{min}}{\Delta p} = \frac{1}{\Delta\bar{p}} \tag{4.40}$$

Note that each logarithm argument in Equation (4.39) can exceed or be less than 1, and that we are considering only integer values for Γ. The upshot is that the terms in the weighted sum can be either positive or negative. Zero is also allowed if, by coincidence, the length fractions match over certain regions. The lower panel of Figure 4.19 shows the results of computing Equation (4.39) for $\Gamma = 100$ states. $KI_{x \leftarrow p}$

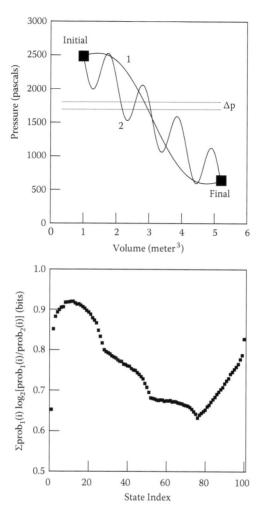

FIGURE 4.19 Reversible pathways and Kullback information. The upper panel shows 1 and 2 having identical pressure ranges. The lower panel shows the summation of Equation (4.39) for $\Gamma = 100$ states. $KI_{X \leftrightarrow p}$ is quantified at approximately 0.80 bits.

is quantified at just over 0.80 bits. The amount is greater at higher resolution, but the qualitative features do not alter. Note that $KI_{X \leftrightarrow p}$ would be infinite if 2 failed outright to anticipate any of the states manifest in 1; one or more logarithm arguments in Equation (4.39) would diverge.

A pathway also expresses mutual information MI_{XY}. This quantifies the pairwise correlation of two control variables, X and Y. One refers to Figure 4.20, which revisits the circle and hairpin. Portions of a grid have been indicated using dotted lines. For each pathway, the probability of observing p and V states in any particular dotted square equates with the length fraction hosted by the square. Such a joint probability differs typically from individual state probabilities. Here, for example, the

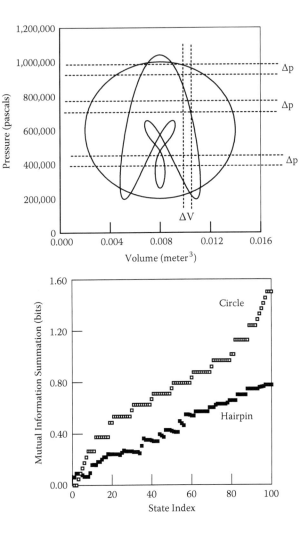

FIGURE 4.20 Reversible pathways and mutual information. The upper panel revisits the circle and hairpin of Figure 4.11. For each pathway, the probability of observing p and V states in any particular square equates with the length fraction contained in the square. The lower panel shows the summation of Equation (4.41) for $\Gamma = 100$ states. $MI_{XY \leftrightarrow pV}$ for the circle is about double that of the hairpin.

probability of observing V states over the ΔV region indicated is given by the sum of two terms. The probability of observing p states over the Δp regions indicated is given by the sum of two to several terms.

$MI_{XY \leftrightarrow pV}$ arrives via the following weighted summation:

$$MI_{XY \leftrightarrow pV} = +\sum_{i,j} prob(\bar{p}_i + \Delta \bar{p}, \bar{V}_j + \Delta \bar{V}) \cdot \log_2 \left[\frac{prob(\bar{p}_i + \Delta \bar{p}, \bar{V}_j + \Delta \bar{V})}{prob(\bar{p}_i + \Delta \bar{p}) \cdot prob(\bar{V}_j + \Delta \bar{V})} \right] \quad (4.41)$$

where $\overline{p}_i + \Delta\overline{p}$, $\overline{V}_i + \Delta\overline{V}$ mark the state widths and boundaries. The number of terms is dictated by both the p and V resolution. As with the Kullback information, the logarithm argument for each term can exceed or be less than 1. Arguments equal to 1 are incompatible with structured pathways. For isotherms and adiabats, $MI_{XY \leftrightarrow pV} > 0$ since the compressibilities β_T and β_S necessarily exceed zero. The lower panel of Figure 4.20 shows the results of applying Equation (4.41) to a 10×10 grid, that is, $\Delta\overline{p} = \Delta\overline{V} = 0.10$. $MI_{XY \leftrightarrow pV}$ for the circle proves to be almost twice that of the hairpin. This reflects the greater correlation of the control variables. For a system programmed for the circle route, a measurement of p tells the chemist more about the V status, compared with hairpin programming.

The major points of Chapter 4 are as follows:

1. A reversible pathway portrays a thermodynamic program applied to a system. The steps are carried out through parallel tuning of the state variables, and exchanges of heat and work coordinated with the surroundings. In each step, a state is converted to a specified neighbor. A locus of states is defined by a program; each state meets the equilibrium criteria described in Chapter 3.
2. For an individual state, fluctuations confer little information in the statistical sense. The information is significantly augmented, however, when structured programs are applied to the system. Thus, their analysis in bit terms offers added perspective of the energy exchanges and variable tuning. The examples of this chapter concentrated on ideal gases and elementary pV transformations. The lessons apply just as readily to more complicated systems, equations of state, and control variables. Information analysis bridges thermodynamic transformations with the probability sciences.
3. More than one type of information is expressed by a program. A pathway can be assessed not only for the Shannon information, but also Kullback and mutual information.

4.5 SOURCES AND FURTHER READING

Transformations figure prominently in thermodynamic texts. Especially recommended is the book by Fermi, which provides a succinct and clear treatment of classical transformations [1]. Adiabatic transformations have widespread applications in science, engineering, and meteorology. The text by Hecht details the thermodynamics of supersonic expansions [2]. Zemansky offers an engrossing account of the high temperatures generated via shock waves. The book is equally instructive about ultracold refrigeration methods [3].

Probability can be approached from multiple vantage points. Penrose offers a penetrating discussion on subjective probability, its foundation, and applications [4].

This chapter addressed pathway length in dimensionless terms. The extensive literature surrounding thermodynamic length with dimensions equivalent to the square root of energy is important; this quantity connects with the work available from a system. The geometrical aspects of thermodynamics have been addressed in papers

by Weinhold [5,6]. Berry, Salamon, and coworkers have established the ramifica-
tions of thermodynamic length beyond equilibrium systems [7–9]. Note that the idea
of length can be quantified in alternative statistical terms. The work of Wootters is
highly instructive on this account [10].

Length arguments figure in diverse thermodynamic applications. The treatise
on small systems by Hill features an application whereby length has dimensions of
$\sqrt{\frac{V}{n} + T}$ [11]. The information properties of classical thermodynamic transformations
have been investigated by the author and student and described in two papers [12,13].
An experimental investigation of information and work costs in chromatographic
systems has also been carried out by the author and students [14].

Last, there is information theory and there is algorithmic information theory. The
reader is encouraged to consult the classic text by Chaitin [15]. Wilf also presents a
rigorous treatment of algorithms and information contexts [16]. Zurek has explored
the subject in detail as well [17,18].

4.6   SUGGESTED EXERCISES

4.1   (a) Let p and V serve as control variables for a thermodynamic pathway of interest. What is the minimum length Λ? (b) Does the answer
change if the identical states are programmed using T and S as control
variables?

4.2   Let 1.00 mole of ideal monatomic gas be transformed along a reversible
pathway by tuning p and V. Let the initial temperature and volume be
400 K and 10^{-3} meter^3, respectively. Let the final volume equal eight
times the initial. Let the final pressure equal one-half of the initial while
the pathway manifests a straight line in the pV plane. (a) Calculate $\langle p \rangle$
and σ_p. (b) Calculate $\langle T \rangle$ and σ_T. (c) Let the chemist query the p
states at resolution equal to 0.50% of the total range. What is $I_{X \leftrightarrow p}$ in
bits? (d) Let the chemist query the temperature states at a resolution
window equal to 0.50% of the total T-range. What is $I_{X \leftrightarrow T}$ in bits? (e)
Let the window for measuring p and V both be 10% of their respective
ranges. What is $MI_{XY \leftrightarrow pV}$ in bits?

4.3   Pathways A and B appeared in several figures of this chapter. They are
detailed by the parametric equations:

$$V_{A,B}(\eta) = 1.00 \text{ meter}^3 \cdot [1 + 1.40\eta]$$

$$p_A(\eta) = 10^3 \text{ pascals} \cdot \left[2.49 - 0.623\eta + 0.374 \sin\left(\frac{2\pi\eta}{3}\right) \right]$$

$$p_B(\eta) = 10^3 \text{ pascals} \cdot \left[2.49 - 0.623\eta + 0.374 \sin\left(\frac{1\pi\eta}{3}\right) \right]$$

where $0 \leq \eta \leq 3$. (a) Calculate σ_p for each pathway. (b) The probability distribution functions regarding pressure appear in Figure 4.13.
Use these to sketch the probability density functions $f_{X \leftrightarrow p}$ for A and

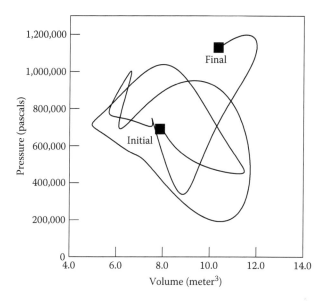

FIGURE 4.21   Reversible pathway in pictorial form only. The plot accompanies Exercise 4.10.

B. (c) Revisit the moment generating function of Chapter 2. For both pathways, construct and sketch $M_{X \leftrightarrow p}(t)$. Use reduced units $\bar{p}$ for the pressure.

4.4 A chemist faxes to a colleague a plot of an interesting pathway. The fax includes the system identity, namely, 1.00 mole of neon gas. No other data are included, however. The plot appears in Figure 4.21. (a) Digitize the pathway as a set of p,V pairs, either by hand or using an optical scanner. (b) Construct $F_{X \leftrightarrow p}$. (c) Estimate values for $\langle p \rangle$ and σ_p. (d) In pressure queries exercised at 2% resolution, what is $I_{X \leftrightarrow p}$ in bits?

4.5 The hairpin pathway was illustrated in Figure 4.11. Let 1.00 mole of ideal monatomic gas travel one circuit. (a) Estimate values for $\langle p \rangle$ and σ_p. (b) What is the maximum efficiency for heat → work conversion?

4.6 Refer to A of Figure 4.6. How many equal-length pathways expressing $I_{X \leftrightarrow S} = 0$ bits can intersect the initial state? Let the same question be directed at the final state. Please discuss in terms of the Carathéodory statement of the second law of thermodynamics. Rigorous discussions of the Carathéodory approach are presented in excellent texts by Chandrasekhar and Reiss [19,20].

4.7 Refer to A in Exercise 3. Let a perturbed version be programmed having the parametric form:

$$V_A(\eta) = 1.00 \text{ meter}^3 \cdot [1 + 1.40\eta]$$

$$p_A(\eta) = 10^3 \text{ pascals} \cdot \left[2.49 - 0.623\eta + 0.374\sin\left(\frac{2\pi\eta}{3}\right) + 0.050\sin\left(\frac{6\pi\eta}{3}\right) \right]$$

(a) Let the system consist of 1.00 mole of a monatomic ideal gas. Does the perturbation impact ΔU, Δp, and ΔV evaluated for the initial and final states? (b) Let the same question apply to $I_{X \leftrightarrow U}$, $I_{X \leftrightarrow p}$, and $I_{X \leftrightarrow V}$. Is a pathway's information conserved upon perturbation?

4.8 Two postulates linked reversible pathways to probability. By an alternate approach, let the chemist imagine a collection of equilibrium systems; each state along a pathway of interest is replicated by one member of the collection. Does this line of thought lead to a probability distribution for the pathway states? Does it lead to the same distribution based on the postulates? Please discuss.

4.9 A chemist considers two cyclic transformations involving 1.00 mole of a monatomic ideal gas. For both, the initial state corresponds to respective V and T of 10^{-3} meter^3 and 400 K. In the first case, an isotherm increases the initial volume by a factor of six. The return steps back to the initial state feature an isobar and an isochore. In the second case, an adiabat increases the initial volume by a factor of six. The link to the initial state is then via an isobar and isochore in that order. (a) Which case offers the better heat $\rightarrow$ work conversion efficiency? (b) Which offers the greater work performed per bit of p state information? (c) Which offers the greater work performed per bit of T state information?

4.10 A chemist considers a reversible pathway involving 1.00 mole of a monatomic ideal gas. The pathway is described by the following parametric equations:

$$p(\eta) = 10^5 \text{ pascals} + 10^4 \text{ pascals} \cdot \sin(\eta)$$

$$V(\eta) = 10^{-3} \text{meter}^3 \cdot (1 + \eta)$$

where $0 \leq \eta \leq 2\pi$. Let the p and V measurement window correspond to 5% of their ranges. What is $MI_{XY \leftrightarrow pV}$ in bits?

4.11 Consider again the pathway of Exercise 4.10. Let a pathway defined by:

$$p(\eta) = 10^5 \text{ pascals} + 10^4 \text{ pascals} \cdot \cos(\eta)$$

$$V(\eta) = 10^{-3} \text{meter}^3 \cdot (1 + \eta)$$

serve as a reference. What is $KI_{X \leftrightarrow p}$, in *bits*? Let the measurement resolution window equal 1% of the total pressure range.

4.12 Figure 4.10 included an isotherm for 1.00 mole of a monatomic ideal gas. Consider this pathway and a straight-line path between the initial and final states. (a) Which pathway expresses the larger $MI_{XY \leftrightarrow pV}$? (b) Let the straight-line path serve as a reference for the isotherm at 1% pressure resolution. What is $KI_{X \leftrightarrow p}$, in bits?

4.13 Refer to Figure 4.20. If the chemist knows the system volume to be in the range marked by the vertical dotted lines, how many bits of information are yielded by a pressure measurement? Consider this question for both the circle and hairpin.

REFERENCES

[1] Fermi, E. 1956. *Thermodynamics*, Dover, New York.

[2] Hecht, C. E. 1990. *Statistical Thermodynamics and Kinetic Theory*, W. H. Freeman, New York.

[3] Zemansky, M. W. 1964. *Temperatures Very Low and Very High*, Dover, New York.

[4] Penrose, O. 2005. *Foundations of Statistical Mechanics*, Dover, New York.

[5] Weinhold, F. 1976. Geometric Representation of Equilibrium Thermodynamics, *Acc. Chem. Res.* 9, 232.

[6] Weinhold, F. 1975. Metric Geometry of Equilibrium Thermodynamics, *J. Chem. Phys.* 63, 2479, and papers that follow.

[7] Salamon, P., Nulton, J. D. 1985. Length in Statistical Thermodynamics, *J. Chem. Phys.* 82, 2433.

[8] Salamon, P., Nulton, J., Ihrig, E. 1984. On the Relation between Entropy and Energy Versions of Thermodynamic Length, *J. Chem. Phys.* 80, 436.

[9] Salamon, P., Berry, R. S. 1983. Thermodynamic Length and Dissipated Availability, *Phys. Rev. Letts.* 51, 1127.

[10] Wootters, W. K. 1980. Statistical Distance and Hilbert Space, *Phys. Rev. D* 23, 357.

[11] Hill, T. 1991. *Thermodynamics of Small Systems*, p. 124 (Part I), Dover, New York.

[12] Graham, D. J. 2009. On the Spectral Entropy of Thermodynamic Paths for Elementary Systems, *Entropy*. DOI: 10.3390/e110x000x.

[13] Graham, D. J., Kim, M. 2008. Information and Classical Thermodynamic Transformations, *J. Phys. Chem. B* 112, 10585.

[14] Graham, D. J., Marlarkey, C., Sevchuk, W. 2008. Experimental Investigation of Information Processing Under Irreversible Brownian Conditions: Work/Time Analysis of Paper Chromatograms, *J. Phys. Chem.* 112, 10594.

[15] Chaitin, G. J. 1987. *Algorithmic Information Theory*, Cambridge University Press, New York.

[16] Wilf, H. S. 1986. *Algorithms and Complexity*, Prentice-Hall, Englewood Cliffs, NJ.

[17] Zurek, W. H. 1989. Algorithmic Randomness and Physical Entropy, *Phys. Rev. A* 40, 4731.

[18] Zurek, W. H. 1984. Reversibility and Stability of Information Processing Systems, *Phys. Rev. Letts.* 53, 391.

[19] Chandrasekhar, S. 1967. *An Introduction to the Study of Stellar Structure*, chap. 1, Dover, New York.

[20] Reiss, H. 1996. *Methods of Thermodynamics*, chap. 4, Dover, New York.

5 State Transformations and Information Economy

Thermodynamic pathways reflect programs that direct a system from one state to another. There are always infinite choices for travel, but not all routes are created equal. Programs offering an economy of length and information are important to multiple fields. This chapter examines the economy issues surrounding reversible pathways.

5.1 DIFFERENT THERMODYNAMIC PATHWAYS WITH IDENTICAL ENDPOINTS

Figure 5.1 illustrates three pathways: A, B, and C. Let all pertain to 1.00 mole of xenon subject to pressure and volume changes. The pathways share initial and final states, and thus demonstrate equivalent changes in state functions: ΔU, ΔH, ΔS, and more. The temperature happens to be the same for the initial and final states, namely, 296 K and thus $\Delta T = 0$. As for all thermodynamic programs, system travel along A, B, or C is driven by variable tuning and energy exchanges. These must be perfectly coordinated to maintain the equilibrium conditions with each step.

The similarities of A, B, and C outnumber the differences. Even so, each presents a unique locus of state points. Pathway A marks an isothermal path that conforms to the van der Waals equation

$$p = \frac{nRT}{V - nb} - \frac{an^2}{V^2} \tag{5.1}$$

where a and b are specific to xenon:

$$a = 0.4236 \, \frac{\text{meter}^6 \cdot \text{pascal}}{\text{mole}^2} \tag{5.2A}$$

$$b = 5.11 \times 10^{-5} \, \frac{\text{meter}^3}{\text{mole}} \tag{5.2B}$$

Pathways A and B are identical in the beginning and terminal regions. Pathway B, however, stems from the Maxwell construction applied to A; the horizontal line divides the loop region into equal areas i and ii. The result is that the gas pressure does not alter over the region bounded approximately by 1×10^{-4} and 4×10^{-4} meter^3. Constant pressure manifests in spite of the volume tuning along B.

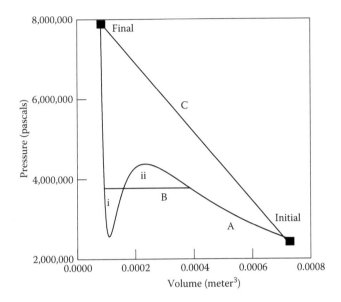

FIGURE 5.1   Pathways with identical initial and final states. A, B, and C pertain to 1.00 mole of xenon subject to pressure and volume changes at fixed temperature.

Pathway C marks a straight-line path in the pV plane. Although the beginning and terminal states share the same temperature, there is a unique T for every intermediate state.

The pathways can be viewed qualitatively. Pathway A is unusual given the loop structure. Over the range $1 \times 10^{-4} - 4 \times 10^{-4}$ meter^3, the pressure falls with decreasing V. How can that be? For a system at equilibrium, the isothermal compressibility

$$\beta_T = \frac{-1}{V}\left(\frac{\partial V}{\partial p}\right)_{T,n} \tag{5.3}$$

must never stray into negative territory. The answer is that Equation (5.1) describing pathway A accounts for some of the nonideality of xenon. The insights arrive, however, with unphysical side effects. One is that β_T is negative for certain combinations of p, V, n, and T. It is the Maxwell construction that serves as an antidote. The horizontal segment of B enables the extraneous work received over one loop of A to be canceled by another.

Pathway C offers the most direct pV route. However, for the xenon to travel C, a large number of heat reservoirs must be switched, one for every temperature state encountered. Regarding pathways A and B, some of the quirks are due to the system undergoing a phase change. Equation (5.1) famously predicts critical point values:

$$T_c = \frac{8a}{27Rb} \tag{5.4A}$$

$$p_c = \frac{a}{27b^2} \tag{5.4B}$$

$$V_c = 3nb \tag{5.4C}$$

These follow from three conditions at the critical point:

$$p_c = \frac{nRT_c}{V_c - nb} - \frac{an^2}{V_c^2} \tag{5.5A}$$

$$\left[\left(\frac{\partial p}{\partial V}\right)_{n,T=T_c}\right]_{V=V_c} = 0 \tag{5.5B}$$

$$\left[\left(\frac{\partial^2 p}{\partial V^2}\right)_{n,T=T_c}\right]_{V=V_c} = 0 \tag{5.5C}$$

The van der Waals model offers fair to decent predictions about gas → liquid transitions. In the case of xenon, the model predicts $T_c \approx 297$ K upon substitution in Equation (5.4A). This compares with the experimental value of 290 K [1]. This correspondence to the experiment is remarkable given the simplicity of Equation (5.1)— only two parameters are used to account for a complicated mix of attractive and repulsive interactions.

As the gas turns into liquid along B, the atoms pack more densely. There are subsequently fewer atoms to collide with the container walls and, in turn, an attached barometer. The volume can indeed be decreased over one portion of the isotherm without effecting a pressure increase. It is only when all the xenon has converted to liquid, at $V \approx 10^{-4}$ meter^3, that the pressure starts to climb again.

Figure 5.1 conveys three programs—an infinite number is possible—that link the designated initial and final states. Not surprisingly, pathways A, B, and C demonstrate some information properties in common and deviate in others. All apply to a closed system. Thus n is constant and zero $I_{X \leftrightarrow n}$ rules the day. Uncertainty would not precede repeated measurements by a chemist able to weigh or count the atoms of the system. $T = 294$ K is allied with all points of A and B; zero $I_{X \leftrightarrow T}$ applies to these pathways. A, B, and C offer nonzero information in multiple variables: p, V, U, S, μ, and more. Energy exchanges and variable tuning underlie all information in the statistical sense.

The methods introduced in Chapter 4 can be directed to A, B, and C to construct probability distributions. Information is one attribute of each distribution of interest. For example, Figure 5.2 shows the probability distributions allied with the pressure states. The curves follow from (1) rescaling the p,V control variables into dimensionless forms that span 0 and 1, (2) computing the contour length fractions

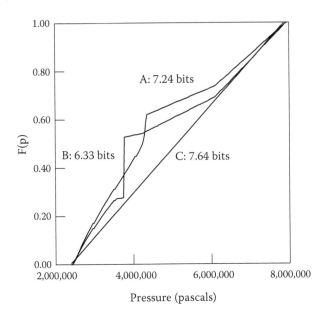

FIGURE 5.2   Pressure state distributions. A, B, and C refer to pathways of Figure 5.1. The Shannon information follows from pressure queries and measurements at resolution equal to 0.50% of the total range.

over each range of states, and (3) converting the reduced variables back into the physical units of choice. It is apparent that all the pressure increments along C manifest equal likelihood. The distribution is uniform as encountered in the first peptide exercise of Chapter 2. If the system is programmed for C travel, and the chemist extends queries such as the chemist anticipates affirmative answers with equal likelihood.

- Does the pressure lie between 4.0×10^6 and 6.0×10^6 pascals?
- Does the pressure lie between 6.0×10^6 and 8.0×10^6 pascals?

The likelihood (probability) would be approximately 0.36 given that each of the ranges are approximately 36% of the total pressure span:

$$\frac{(6.0 - 4.0) \times 10^6 \text{ pascals}}{p_{max} - p_{min}} \approx 0.36 \tag{5.6A}$$

$$\frac{(8.0 - 6.0) \times 10^6 \text{ pascals}}{p_{max} - p_{min}} \approx 0.36 \tag{5.6B}$$

The A and B distributions are skewed and, as a consequence, their pressure states pose less uncertainty prior to measurement. This is reflected in $I_{X \leftrightarrow p}$. Taking Δp to be

0.50% of the total range, that is, $\Delta p \approx 2.8 \times 10^4$ pascals, $I_{X \leftrightarrow p}$ for C becomes

$$
\begin{aligned}
I_{X \leftrightarrow p} &= -\sum_{j=1}^{200} prob(j) \cdot \log_2(prob(j)) \\[2mm]
&= \frac{-1}{\log_e(2)} \sum_{j=1}^{200} prob(j) \cdot \log_e(prob(j)) \\[2mm]
&= \frac{-1}{\log_e(2)} \cdot 200 \times \frac{1}{200} \cdot \log_e\left(\frac{1}{200}\right) \\[2mm]
&= \frac{+1}{\log_e(2)} \cdot \log_e(200) \approx 7.64 \text{ bits}
\end{aligned}
\tag{5.7}
$$

$I_{X \leftrightarrow p}$ for A and B follow from applying Equations (5.28) and (5.30) using identical Δp ($\approx 2.8 \times 10^4$ pascals), the results being 7.24 and 6.33 bits, respectively. One learns that while C traces the most direct route in the pV plane, it is the most expensive of the three in the code needed for labeling the pressure states. The van der Waals isotherm A places second in code expenditures. Pathway B, which most accurately portrays xenon in real life, proves the most economical in labeling costs. Not incidentally, in labeling the temperature states, C's price tag for code is infinitely greater than A's or B's.

Elementary pathways offer a reprise of Chapter 4 and introduce the next topic. Figure 5.3 goes one step further. Given two states of a system, there are an unlimited number of travel recipes ranging from the simplest and most direct to the complicated and meandering. Different pathways express different amounts of information and, in turn, labeling costs in query-and-measurement exercises. Whereas Chapter 4 focused on the methods of quantification, Chapter 5 looks at issues of economy. It addresses the question: What are the most frugal programs in terms of length and information that can link two states of interest? Figure 5.3 portrays points located by generic state variables X and Y. What path should be programmed to join the initial and final with the minimum code expenditures? Paths 1 and 2 both seem needlessly lengthy and complicated. How should "?" be constructed for maximum economy? A lesson of Figures 5.1 and 5.2 is that the shortest route does not necessarily offer the information bargain.

The topic brings to mind applications beyond gases and pV tuning. This is because the challenge of transforming something from one state to another, be it material or abstract, is ubiquitous. Trains need to travel from here to there; crews seek and construct the best track routes. Computational tasks—sorting, addition, and so forth—convert one set of bytes to another. Programmers develop and code the best algorithms. Chemists convert stockroom reagents into natural products. They design and execute the best synthetic routes. Yet, terms like *best route* warrant further exposition. This is provided in examples drawn from computation and chemistry.

Consider the process of integer factoring, which impacts modern computer security. It is well appreciated that an integer N is either prime or composite. If prime, N

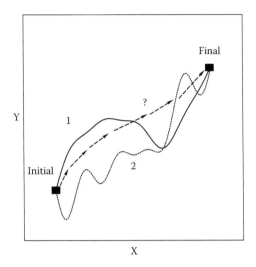

FIGURE 5.3 Pathways and economy. There are unlimited ways to connect initial and final states by tuning the control variables X and Y. What are the most strategic pathways in length and information?

can be divided without remainder only by itself and 1; 2, 3, 5, 7, 11, 13, ... constitute the prime number series. If N is composite, it can be expressed as a unique product of primes, for example:

$$4 = 2 \times 2$$

$$6 = 2 \times 3$$

$$100 = 2 \times 2 \times 5 \times 5$$

There are an infinite number of composite integers and the same is true for primes.

Integers do not factor by themselves. Each requires stepwise conversion along a specified path. The transistor circuits of computers execute the conversions in daily life. The work and heat exchanges that drive them are controlled by factoring programs.

Clearly, some paths are more strategic than others. A prime factor of N cannot exceed $N^{1/2}$. It makes little sense, however, to commence a factoring job by searching for and testing primes near $N^{1/2}$. This is because the density of primes scales as $1/\log_e(N)$. For arbitrary N, it is more likely that low-valued integers have prime status *and* prove to be viable factors. After all, 50% of integers have 2 as a prime factor.

$N = 159{,}870{,}864{,}030$ presents a modest challenge with prime factors 2, 3, 5, 547, 1229, and 7927. In arriving at these by trial and error, the steps needed for $N^{1/2}$-downward travel far exceed those of a 2-upward path. Each step requires processing 32 bits or more in the typical laptop computer. Both top-down and bottom-up procedures take 159,870,864,030 to the same final state: $2 \times 3 \times 5 \times 547 \times 1229 \times 7927$ or equivalent. The economical route, however, navigates through the states of greatest fractional occurrence—the low-value integers that are most likely to be prime numbers and factors of N.

Pathway economy is no less important in chemistry. Peptides were used to illustrate the probability functions of Chapter 2. A different question is raised here: If a chemist needed to prepare a polypeptide from scratch—and with no help from cellular machinery—what strategy should he or she elect?

Here the initial state consists of the 20 amino acids kept in stockroom bottles: A, V, G, and so on. The final state is an N-unit peptide such as GLVDAKNDVAR … WHSV. As with factoring, there is more than one road to travel, some much less appealing than others. For $N = 256$, a *serial* procedure would be ill-advised. Even if each step transpired at 95% yield, the outcome would be withering given that $(0.95)^{255} \approx 2.1 \times 10^{-6}$. If the amino acids are transformed along this route, the chemist ends up taking 1.00 mole of starting reagent G to 2 micromoles of GLVDAKNDVAR … WHSV.

Convergent designs are superior by far. Matters are initiated by the preparation of dipeptides GL, VD, AK, ND, … WH, SV—128 total. The chemist links these to make tetrapeptides GLVD, AKND, …, WHSV. The tetrapeptides are then transformed to octapeptides and so on. The N-unit peptide thereby requires $\log_2(N)$ stages of assembly. If the individual yields are 95%, the final yield is $(0.95)^8 \approx 0.663$. The initial and final states are the same in both procedures: amino acids in bottles and GLVDAKNDVAR … WHSV. Yet the second route is favored yieldwise by five orders of magnitude. The chemist makes frugal use of the reagents, solvents, chromatography supplies, and so forth.

Note the strategies practiced by factoring and peptide synthesis. Both process rightly chosen building blocks. Factoring programs do not waste time and energy with floating point variables: 547 is tested as a prime factor, not 547.0000000000. Chemists do not try to steer ethane molecules toward GLVDAKNDVAR … WHSV. Both enterprises target the shortest pathways with no meandering. If a program verifies 2, 3, 5, and 547 as prime factors, it does not stray by examining the sum $2 + 3 + 5 + 547$. When the chemist reaches the octapeptide stage, the molecules are not subjected to unnecessary side reactions.

It is important that integer factoring and synthesis walk through the states of greatest likelihood: the most likely prime candidates and the highest-yield (i.e., most frequent) intermediates. Information underlying heat and work exchanges makes both endeavors feasible. The reader is directed to the end-of-chapter references on high-throughput synthesis, linear programming, and computational mathematics. Identifying the optimum procedures is a never-ending challenge in these and companion fields.

Pathway design is critical in thermodynamic applications. These include heat engines, distillation, refrigeration, and petroleum cracking, to name a few. Regarding the first, Sadi Carnot inquired about the pathways that best convert heat into work. Cyclic pathways—the programs for heat engines—were featured in Figures 4.11, 4.15, and 4.20 in Chapter 4. For the discussion at hand, Figure 5.4 presents one example of a Carnot pathway marked by the solid curves. The transformation processes 1.00 mole of a monatomic ideal gas over the temperature range 400 to 800 K. The cycle has been illustrated in the pV plane for convenience, although other coordinate planes are equally instructive. The feature to note is that the gas is transformed in four stages: i, ii, iii, and iv label the endpoints of two isotherms (upper and lower)

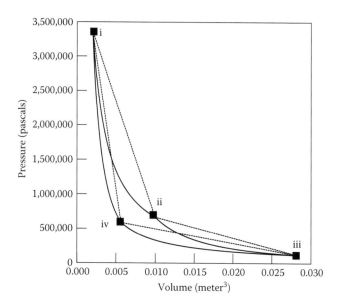

FIGURE 5.4 Carnot and alternative pathways. The solid curves trace two isotherms (800 and 400 K) and two adiabats for 1.00 mole of ideal gas. The dotted pathway marks one of infinite cyclic transformations that share the same pV boundaries as the Carnot.

and two adiabats (left and right). Each segment is distinguished by its placement and steepness in the pV plane. The 800 and 400 K isotherms have respective endpoints of {i, ii} and {iii, iv}. The adiabats have endpoints {ii, iii} and {i, iv}. For the system to travel along the pathway, there are no restrictions on the initial state or number of circuits. Thus, the initial state does not have to correspond to i, ii, iii, or iv; it can be anywhere on any segment. What is vital is that the initial and final state be identical. If the transit is clockwise, heat is injected from an external reservoir and converted partially to work. For counterclockwise travel, heat is pumped from a cooler to hotter reservoir at a cost of externally supplied work. A Carnot cycle provides the most famous, if idealized, model for heat engines and refrigerators.

The dotted lines in Figure 5.4 trace one of infinite transformations that fall within the Carnot pV boundaries. Four straight-line segments demonstrate the same endpoints as the isotherms and adiabats. The appearance is that of a bent diamond or perhaps a boomerang. The question is raised: If the Carnot or diamond path is traversed clockwise by 1.00 mole of ideal gas, which offers the more effective conversion of heat into work?

Isothermal and adiabatic pathways were discussed in Section 4.2 of Chapter 4. One learned that the following statements hold for the isotherms of Figure 5.4:

$$p \cdot V = nRT = Constant \tag{5.8}$$

$$\Delta U = C_V \cdot \Delta T = 0 \tag{5.9}$$

Relating Equation (5.9) to the first law of thermodynamics, it follows that

$$Q_{rec} = -W_{rec} = +W_{performed} \tag{5.10}$$

for upper and lower isotherms. For each adiabat, the following statements apply:

$$p \cdot V^{\gamma} = Constant \tag{5.11}$$

where

$$\gamma = \frac{C_p}{C_V} \tag{5.12}$$

and

$$Q_{rec} = \int T \, dS = 0 \tag{5.13}$$

In connecting Equation (5.13) to the first law, one obtains

$$\Delta U = +W_{rec} \tag{5.14}$$

for the left and right adiabats.

Equations (5.8) through (5.14) identify the signature features of a Carnot program for an ideal gas. The finer points are noted as follows:

1. Work is exchanged between the gas and surroundings along each segment. The heat exchanges are confined to the isotherms, however.
2. The work lost by the system along the {ii, iii} adiabat is restored along the {i, iv} adiabat.
3. The overall (total) work performed with every circuit equates with the area enclosed by the segments. Since

$$\Delta U = 0 \tag{5.15}$$

regardless of where travel commences, then

$$W_{rec}^{(total)} = -Q_{rec}^{(total)} \tag{5.16}$$

which reflects the first law impact yet again.

4. The Carnot efficiency ε equates with the ratio:

$$\varepsilon = \frac{Total\ Work\ Performed}{Heat\ Injected\ along\ Upper\ Isotherm} \tag{5.17}$$

In spite of the nontrivial structure of the pathway, ε acquires a most compact form for an ideal gas:

$$\varepsilon = \frac{T_{max} - T_{min}}{T_{max}}$$

$$= 1 - \frac{T_{min}}{T_{max}} \tag{5.18}$$

Thus, for the Carnot cycle of Figure 5.4, ε is computed as

$$\varepsilon = 1 - \frac{400\ K}{800\ K} = 0.500 \tag{5.19}$$

Clearly, the efficiency would be higher if the reservoir temperatures were further apart. Note importantly that ε does not alter if the pV boundaries are changed (e.g., the isotherms are lengthened or shortened), or if more or less gas composes the system.

It is instructive to compare the results for the diamond. Here W_{rec}, Q_{rec}, and ΔU must be evaluated over each of the four stages. W_{rec} for a stage equates with the area underneath each straight-line segment—a combination of triangle and rectangle shapes. As with any cyclic path, total W_{perf} equals the total enclosed area. The state conversions with positive Q_{rec} identify where heat is injected, whereas those with negative W_{rec} quantify the output work. The heat $\rightarrow$ work efficiency follows by dividing the total output work by the injected heat. The final expression for the efficiency is not compact as with a Carnot cycle. The diamond efficiency can be shown (Exercise 11) to be just under 30%. This is significantly less than that of Equation (5.19) in spite of the wider temperature range: 400 to 1000 K. The efficiency of a Carnot cycle would be 0.60 if afforded this range.

Carnot cycles are instructive because they illuminate the optimum programs for converting heat into work. And not incidentally, such pathways are information-strategic in temperature and entropy, not to mention moles of material. Every Carnot segment of Figure 5.4 poses nonzero $I_{X \leftrightarrow p}$ and $I_{X \leftrightarrow V}$; likewise for the diamond. Yet, along each isotherm and adiabat, $I_{X \leftrightarrow T}$ and $I_{X \leftrightarrow S}$ equate, respectively, with zero. In contrast, each segment of the diamond expresses virtually maximum $I_{X \leftrightarrow T}$ and $I_{X \leftrightarrow S}$. This is because unique T, S pairings attach to every state point, the only exceptions appearing at the endpoints. The temperature and entropy states along a straight-line segment manifest equal likelihood because there is vanishing bias in their distribution. They pose near-maximum uncertainty in measurements directed at the system.

The pathway differences receive further attention in Figure 5.5. Shown are the probability distributions constructed for the temperature states. The results reflect, unsurprisingly, that the temperature distribution is considerably skewed for the Carnot process. A system so programmed would pose less uncertainty in query-and-measurement exercises. The bias is marked because T is at its minimum or maximum value for sizable portions of the cycle. By contrast, the diamond path reflects a greater range and more even dispersion of temperature states. The information values in Figure 5.5 apply

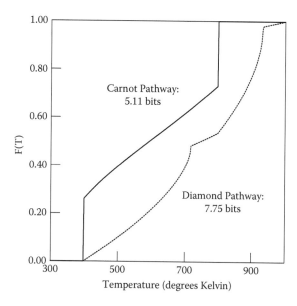

FIGURE 5.5 Probability distributions for temperature states. Distributions apply to the Carnot and diamond pathways of Figure 5.4. The information values follow from query-and-measurement exercises at resolution $\Delta T = 2$ K.

to resolution $\Delta T = 2$ K. It is notable that the diamond incurs approximately 50% greater labeling costs concerning temperature states. The greater expenditure of code does not purchase a more favorable heat → work conversion. Rather, the work returns on heat investments depend critically on the pathway structure. As Carnot paths demonstrate, the choices of information-strategic routes need not be at odds with thermodynamic goals. The next section offers lessons about pathway economy.

5.2 PATHWAY PROGRAMMING AND ECONOMY

There are infinite pathways that can join two states. The first lesson about the programs that determine pathway structure is that they are, in some ways, superfluous. This is surprising. For a closed system to transit from a designated initial state to a final one, the system and surroundings need only to exercise random energy exchanges. Each transaction will reposition the state point. The location may be new or previously visited. If the exchanges are random, there will be no bias shown toward one state or another—they will be visited with equal opportunity. With sufficient exchanges, the pathway will trace out a Brownian pattern whereupon the state point will sample all possible positions, the final one included. Brownian pathways fill space in one and two dimensions. This means that the energy exchanges can take a closed, $\kappa = 1$ system to the final state, expressing myriad p and V along the way. Matters are different if the system is open and n is no longer fixed. Here, not every state will be visited during random exchanges. A structured program becomes mandatory.

An example is shown in Figure 5.6. One considers a single-component gas in a leak-proof container with initial p and V near 50,000 pascals and 0.100 meter³. For

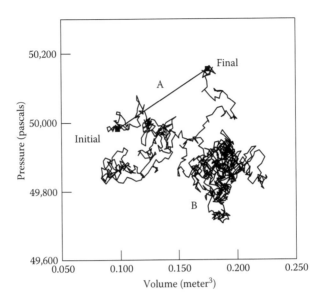

FIGURE 5.6   Straight-line and random pathways. A is the shortest pV route between the initial and final states. B results from random energy exchanges between the system and surroundings.

the system to reach the final state, the surroundings need to inject heat and extract work incrementally. The straight-line path A will certainly do the job.

Random state sampling, however, offers alternatives such as B. Note that A and B are identical in the net changes of all state functions. They contrast in Q_{rec}, W_{rec}, and information in all variables except n. The disparities are portrayed in Figure 5.7 showing the pressure–state probability curves. There is zero bias in the A distribution—the distribution is uniform. Not so for B: roughly 80% of the path accounts for only 50% of the states. During measurements on the chemist's part, the bias diminishes the uncertainty surrounding the states along B. Yet compared with A, the pressure range is greater for B by more than a factor of 2. In quantifying $I_{X \leftrightarrow p}$, there are more weighted surprisal terms to sum for the B distribution. $I_{X \leftrightarrow p}$ quoted in Figure 5.6 follow from Δp set at 0.50% of the range for A: $\Delta p \approx 0.865$ pascals. There are then 200 states for which to allocate code. The Shannon information for A works out to be:

$$I^{(A)}_{X \leftrightarrow p} = \frac{-1}{\log_e(2)} \cdot \sum_{j=1}^{200} prob(j) \cdot \log_e(prob(j))$$

$$= \frac{-1}{\log_e(2)} \cdot 200 \times \frac{1}{200} \times \log_e\left(\frac{1}{200}\right) \qquad (5.20)$$

$$= \frac{+\log_e(200)}{\log_e(2)} \approx 7.64 \text{ bits}$$

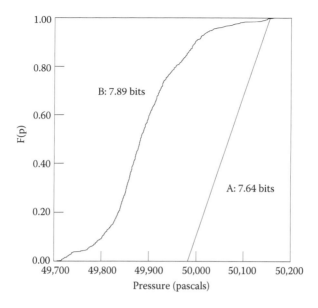

FIGURE 5.7 Probability distributions and pressure states. Distributions apply to the straight-line and random paths of Figure 5.6. $I_{X \leftrightarrow p}$ follow from measurements with Δp set at 0.50% of the range in A: $\Delta p \approx 0.865$ pascals.

At the same physical resolution, there are more than 500 states to accommodate for B and $I_{X \leftrightarrow p}$ works out to be just under 8 bits. Greater information applies in spite of the heavy concentration of states near 49,800 pascals.

It should be apparent why B, while interesting, is a contrived example. With random energy exchanges, there is no guarantee that the state point will relocate in the desired way. It may take infinite steps to reach the wished-for state. A structured program may be superfluous in some respects. However, the lack of one, barring dumb luck, puts the labeling costs as high as infinite bits. This holds for all variables except for n. Simply stated, the programs for structured pathways are vital because they offer economy of both length and information.

A second lesson follows. Shorter pathways generally offer better economy than longer ones. Figure 5.8 illustrates two cases where the initial and final states are shared. Pathway 1 follows a wobbly route from the initial to final state; pathway 2 elects a smoother but longer path. It is not difficult to see which poses the greater uncertainty in pressure state queries. The results of analysis appear in Figure 5.9. The diversity of pressure states in 2 are reflected in the larger $I_{X \leftrightarrow p}$: 8.56 bits versus 7.41 bits for pathway 1. For this calculation, Δp was taken to be 0.50% of the range demonstrated by 1: $\Delta p \approx 0.935$ pascals.

But therein lies a third lesson: the shortest route is not necessarily the most economical. Pathway 1 is wobbly. If the program had opted for a straight line, $I_{X \leftrightarrow p}$ would have exceeded 7.41 bits. Since all states would then have expressed equal likelihood, $I_{X \leftrightarrow p}$ would have equated with the Equation (5.20) result. The wobbles in pathway 1 increase the length. Yet they bring about a distribution bias that enhances the information economy.

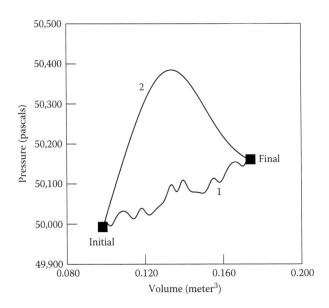

FIGURE 5.8   Shorter and longer pathways. Pathway 1 follows a wobbly route from the initial to final state. Pathway 2 elects a smoother but longer path.

Note the qualifier "not necessarily." Sometimes the most direct pathway *is* the most economical. More to the point, the special pathways of Chapter 4 offer the maximum economy in one or more state variables. For isobaric, isochoric, adiabatic, and isothermal cases, $I_{X \leftrightarrow p}$, $I_{X \leftrightarrow V}$, $I_{X \leftrightarrow S}$, and $I_{X \leftrightarrow T} \approx 0$, respectively; for every path of a closed system, $I_{X \leftrightarrow n} = 0$. Note that special pathways afford straight-line representations in select planes, for example, isobaric and isochoric in pV, and adiabatic in the TS plane. A Carnot cycle appears as a square or rectangle when drawn in the TS plane.

Pathways embody programs for state transformations. The strategic designs aim for the shortest and surest routes. This is the case for integer factoring, organic synthesis, and hopefully railroad construction. Figure 5.10 offers several choices of pV programs. Which offers the most favorable economy?

One considers the five pathways that begin and terminate identically. They vary in length, whereby A marks the shortest route, and E is the longest. All except A express bias in the pressure states. More than 10% of D, for instance, hovers near 50,175 pascals. As for E, a significant fraction threads pressure states identical to the final state.

In query-and-measurement exercises by the chemist, B, C, D, and E afford less uncertainty about the system pressure compared with A; ditto for the volume states. The diminished uncertainty arrives, however, at the expense of programming longer pathways—of having to specify and reckon with more extended collections of state points. Clearly, the strategic designs aim at a favorable trade-off between length and state bias. How should the chemist weigh one program against another on economy grounds?

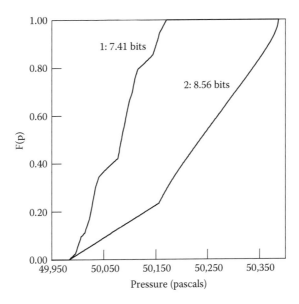

FIGURE 5.9 Probability distributions and pressure states. 1 and 2 pertain to the pathways of Figure 5.8. The diversity of states in pathway 2 are reflected in the larger $I_{X \leftrightarrow p}$: 8.56 bits versus 7.41 bits for pathway 1. Δp was taken to be 0.50% of the range demonstrated by pathway 1: $\Delta p \approx 0.935$ pascals.

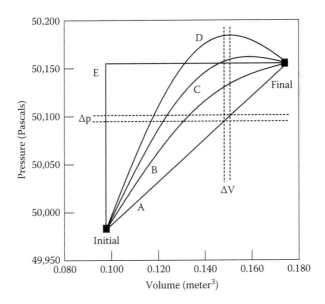

FIGURE 5.10 Assorted pV programs connecting initial and final states. Which offers the best economy in length and information?

Heat engines are evaluated by their efficiency ε; refrigerators are evaluated by their coefficient of performance. When discriminating pathways for length and information economy, a merit parameter Ω_X proves highly useful:

$$\Omega_X = \frac{\Lambda}{\Lambda_o} \cdot \frac{I_X}{I_{X_o}} \tag{5.21}$$

Λ is the reduced (i.e., dimensionless) length of the pathway under consideration (cf. Equations 4.28 and 4.29). Λ_o is the reduced length of the straight-line path connecting initial and final states. I_X is the Shannon information concerning state variable X ($\leftrightarrow p, V, T$, etc.) queried at resolution window ΔX. I_{X_o} is the information expressed by X for the straight-line path queried at the same resolution.

The strategic routes are shorter than ill-designed ones. They are less diverse as well, since the chemist requires fewer bits of code for state labeling. Accordingly, the sought-after pathways demonstrate smaller Ω_X on account of length *and* information. A path with $\Omega_X = 1.75$ offers a better deal overall than one having $\Omega_X = 2.75$. Note, however, that a pathway with $\Omega_X > 1$ is worse than a straight line. It may offer lower information in the probability distribution, but the size of the state point population (i.e., pathway length) that must be programmed cancels any benefits over the simplest and most direct route. By the same token, a pathway with $\Omega_X \approx 1$ offers no substantial improvement over the shortest and simplest route.

The optimum pathways offer minimum Ω_X. Strategic programming looks for ways to reduce Ω_X. A pathway with the minimum Ω_X marks the shortest and surest route, the length and information weighted equally. It is apparent that random or Brownian paths are undesirable at once on account of large Λ/Λ_o ratios. B in Figure 5.6, for example, turns out to have to have $\Lambda \approx 31.3$. Using the information values of Figure 5.7, $\Omega_{X \leftrightarrow p}$ is obtained as

$$\Omega_{X \leftrightarrow p} = \frac{31.3}{\sqrt{2}} \cdot \frac{7.89 \, bits}{7.64 \, bits} \approx 22.1 \tag{5.22}$$

Recall that this applies to a contrived case. Pathways governed by random energy exchanges pose $\Omega_{X \leftrightarrow p}$ as high as infinity.

It is also apparent that isotherms, isobars, isochores, and adiabats should be termed *perfect*. Each presents a thermodynamic variable X for which $I_X = 0$ and likewise for Ω_X. As would be expected, perfect pathways are admitted only by atypical circumstances. Perfect pathways can take a system from initial to final states only when these states happen to share identical X values.

For the typical circumstances, the most economical pathways for linking two state points can be termed *ideal* ones. A and E of Figure 5.10 fall into this category; the choice dependent on the measurement resolution employed by the chemist. E is a dual-segment route that is longer than straight-line A. Statewise, however, it is the more certain route because one variable, in addition to n, is held constant along each segment. The computation of $\Omega_{X \leftrightarrow p}$ for E illustrates matters.

A and E of Figure 5.10 share initial and final states. Because of its straight-line nature, A serves as the yardstick for evaluating E. The pressure and volume ranges are identical for A and E. Toward computing $\Omega_{X \leftrightarrow p}$, one rescales the control variables of A and E:

$$\bar{p} = \frac{p - p_{min}}{p_{max} - p_{min}} \tag{5.23A}$$

$$\bar{V} = \frac{V - V_{min}}{V_{max} - V_{min}} \tag{5.23B}$$

The results range from 0 to 1 whereby rescaled A and E have been replotted in Figure 5.11. The reduced length along each E segment is 1, hence, Λ is $1 + 1 = 2$. The analysis applied to A gives $\Lambda_o = 2^{1/2}$. Two of the slots in the expression for $\Omega_{X \leftrightarrow p}$ become filled immediately, namely,

$$\frac{\Lambda}{\Lambda_o} = \frac{2}{\sqrt{2}} = \sqrt{2} \approx 1.41 \tag{5.24}$$

One then turns to the remaining slots. In query-and-measurement exercises, the number of discernable states $\Gamma_{X \leftrightarrow p}$ is set by the resolution

$$\Gamma_{X \leftrightarrow p} = \frac{p_{max} - p_{min}}{\Delta p} = \frac{1}{\Delta \bar{p}} \tag{5.25}$$

For simplicity of illustration, the resolution Δp has been taken to be one-fifth of the pressure range of A and E. Thus $\Gamma_{X \leftrightarrow p} = 5$, and the states have been indexed accordingly in Figure 5.11.

Now suppose that the chemist tendered queries about a system programmed for E travel. Valid questions would include: What is the likelihood that the pressure, upon measurement, falls in the range defined and labeled as state 1, that is, close to maximum p? It should be clear that half of the state points lies along the horizontal segment affiliated with p_{max}. One-half lies along the vertical as well, but only one-fifth of these fall into the region designated as state 1. Thus, the chemist arrives at a reasonable-belief probability of

$$prob(1) = \frac{1}{2} + \frac{1}{2 \times 5} \tag{5.26}$$

Just as valid a question is: What is the likelihood that the pressure corresponds to state 5? The chemist reasons that there is a 50% chance that the state point lies along the vertical; 20% of the vertical points lie in the bottom fifth. Thus, the answer would be:

$$prob(5) = \frac{1}{2} \times \frac{1}{5} \tag{5.27}$$

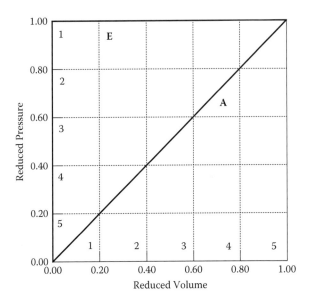

FIGURE 5.11 Rescaled A and E of the previous figure. The reduced length Λ is $1 + 1 = 2$ for E; $\Lambda = 2^{1/2}$ for A. The dotted grid marks the pressure and volume states discussed in the text. The integers label the pressure and volume states whereby $\Delta \overline{V} = \Delta \overline{p} = 0.200$.

Clearly, E demonstrates a marked bias; half of the pressure states are indeed pooled at one value. The equivalent statement holds as well for the volume states. Note the contrast with A. The answers to the same questions posed for a system programmed for A travel would be equal at one-fifth. In effect, E offers less uncertainty about the system pressure, but at the expense of greater program length. To decide matters of economy, there are two more slots to fill for $\Omega_{X \leftrightarrow p}$.

Realizing the denominator $I_{X_o \leftrightarrow p}$ is straightforward; it is the pressure information affiliated with A:

$$I^{(A)}_{X \leftrightarrow p} = \frac{-1}{\log_e(2)} \cdot \sum_{j=1}^{5} prob(j) \cdot \log_e(prob(j))$$

$$= \frac{-1}{\log_e(2)} \cdot 5 \times \frac{1}{5} \times \log_e\left(\frac{1}{5}\right) \tag{5.28}$$

$$= \frac{+\log_e(5)}{\log_e(2)} \approx 2.32 \text{ bits}$$

Obtaining numerator $I^{(E)}_{X \leftrightarrow p}$ is more involved:

$$I^{(E)}_{X \leftrightarrow p} = \frac{-1}{\log_e(2)} \cdot \sum_{j=1}^{5} prob(j) \cdot \log_e(prob(j))$$

$$= \frac{-1}{\log_e(2)} \cdot \left\{ \left(\frac{1}{2} + \frac{1}{2 \times 5} \right) \cdot \log_e \left(\frac{1}{2} + \frac{1}{2 \times 5} \right) + \sum_{j=2}^{5} prob(j) \cdot \log_e(prob(j)) \right\}$$

(5.29)

$$= \frac{-1}{\log_e(2)} \cdot \left\{ \left(\frac{1}{2} + \frac{1}{2 \times 5} \right) \cdot \log_e \left(\frac{1}{2} + \frac{1}{2 \times 5} \right) + (5-1) \times \left(\frac{1}{2 \times 5} \right) \cdot \log_e \left(\frac{1}{2 \times 5} \right) \right\}$$

One arrives at $\Omega_{X \leftrightarrow p}$ by combining Equations (5.24), (5.28), and (5.29):

$$\Omega_{X \leftrightarrow p} = \frac{2}{\sqrt{2}} \cdot \frac{1.77 \text{ bits}}{2.32 \text{ bits}} \approx 1.08$$

(5.30)

Note the important result: E has proven less favorable than A. Economy of length has trumped the economy of information! If an identical twin of A had been considered (i.e., A referenced against itself), the results would have been $\Omega_{X \leftrightarrow p} = 1$. But at the same time, note how matters change with the query resolution. If the chemist were able to discern, say, 50 pressure states instead of 5, one obtains:

$$I_{X \leftrightarrow p}^{(E)} = \frac{-1}{\log_e(2)} \cdot \sum_{j=1}^{50} prob(j) \cdot \log_e(prob(j))$$

$$= \frac{-1}{\log_e(2)} \cdot \left\{ \left(\frac{1}{2} + \frac{1}{2 \times 50} \right) \cdot \log_e \left(\frac{1}{2} + \frac{1}{2 \times 50} \right) + (50-1) \times \left(\frac{1}{2 \times 50} \right) \cdot \log_e \left(\frac{1}{2 \times 50} \right) \right\}$$

$$\approx 3.75 \text{ bits}$$

(5.31)

whereupon

$$\Omega_{X \leftrightarrow p} = \frac{2}{\sqrt{2}} \cdot \frac{3.75 \text{ bits}}{\frac{1}{\log_e(2)} \cdot \log_e(50)} \approx 0.939$$

(5.32)

Thus, at the higher query resolution, E offers distinct advantages over the shortest programming route A. In particular, the bias in the state distribution more than compensates for the greater length.

Importantly, Equation (5.31) can be generalized for Γ_X number of states; X can represent any control variable V, T, and so forth:

$$I_X = \frac{-1}{\log_e(2)} \cdot \sum_{j=1}^{\Gamma_X} prob(j) \cdot \log_e(prob(j))$$

$$= \frac{-1}{\log_e(2)} \cdot \left\{ \left(\frac{1}{2} + \frac{1}{2\Gamma_X} \right) \cdot \log_e \left(\frac{1}{2} + \frac{1}{2\Gamma_X} \right) + (\Gamma_X - 1) \times \left(\frac{1}{2\Gamma_X} \right) \cdot \log_e \left(\frac{1}{2\Gamma_X} \right) \right\}$$

(5.33)

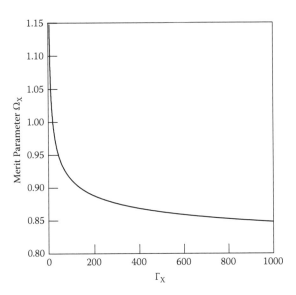

FIGURE 5.12   The dependence of Ω_X on Γ_X. The plot shows that for $\Gamma_X \leq 16$, the straight-line path A offers the more favorable economy. For $\Gamma_X > 16$, the dual-segment pathway E offers the more favorable Ω_X. The merit parameter levels off slowly with increasing Γ_X.

Figure 5.12 shows the dependence of Ω_X on Γ_X. The plot illustrates that for $\Gamma_X \leq 16$, the straight-line path offers the greater programming economy. Things are otherwise for $\Gamma_X > 16$ where a dual-segment pathway offers the more favorable Ω_X. At the higher resolution, the ideal pathway is simply a combination of two perfect pathways. Figure 5.12 also shows that the merit parameter levels off—albeit slowly—with increasing Γ_X. One can identify the lower bound for the ideal pathway by approximating Ω_X at very large Γ_X:

$$\Omega_X = \frac{\Lambda}{\Lambda_o} \cdot \frac{I_X}{I_{X_o}}$$

$$= \left[\frac{2}{\sqrt{2}} \right] \cdot \left[\frac{\left[-1 \times \left\{ \left(\frac{1}{2} + \frac{1}{2\Gamma_X} \right) \cdot \log_e \left(\frac{1}{2} + \frac{1}{2\Gamma_X} \right) + (\Gamma_X - 1) \cdot \left(\frac{1}{2\Gamma_X} \right) \cdot \log_e \left(\frac{1}{2\Gamma_X} \right) \right\} \right]}{\log_e(2)} \right] \qquad (5.34)$$

$$\approx \left[\frac{2}{\sqrt{2}} \right] \cdot \frac{-1 \times \left\{ \left(\frac{1}{2} \right) \cdot \log_e \left(\frac{1}{2} \right) + \left(\frac{1}{2} \right) \cdot \log_e \left(\frac{1}{2} \right) + \left(\frac{1}{2} \right) \cdot \log_e \left(\frac{1}{\Gamma_X} \right) \right\}}{\log_e(\Gamma_X)}$$

$$\approx \left[\frac{2}{\sqrt{2}} \right] \cdot \frac{+ \left\{ \log_e(2) + \left(\frac{1}{2} \right) \cdot \log_e(\Gamma_X) \right\}}{\log_e(\Gamma_X)}$$

$$\approx \left[\frac{2}{\sqrt{2}} \right] \cdot \frac{1}{2} = \frac{1}{\sqrt{2}} \approx 0.707$$

The result shows that the ideal pathway has a merit parameter that exceeds zero and is thereby less than perfect. Yet there is a limit to the ideality: Ω_X cannot drop below 0.707 given appreciable measurement resolution.

5.3 PROPERTIES OF PATHWAY LENGTH AND INFORMATION ECONOMY

Chemical thermodynamics offers ways to categorize and model systems: ideal, non-ideal, closed, open, and so forth. The subject does likewise for pathways along which a system is transformed: reversible versus irreversible, isobaric, isochoric, cyclic, and more. The focus of this chapter has been programming economy—the criteria being pathway length and information. Section 5.2 pointed to four categories along strategy lines. A reversible pathway falls into one of the following:

1. A pathway that is economically perfect in length and information terms offers a control variable X such that $\Omega_X = 0$. For a closed, single-component system, X and n are fixed. The specification of only one other variable Y is needed for locating every state point along the pathway.
2. A worst-case pathway is one absent of a thermodynamic algorithm. The state point placements demonstrate a Brownian nature, whereby Ω_X can be as high as infinity on length and information accounts.
3. An ideal pathway is one where, for control variable X, $0.707 \leq \Omega_X \leq 1$. Atypical conditions enable transformations along perfect pathways. Typical conditions, by contrast, always admit system programming via ideal pathways.
4. Pathways for which none of the above applies are less than ideal. For control variable X, one has $1 < \Omega_X \ll \infty$.

There are additional properties to note. First, the economy does not depend on the direction of travel. Regardless of which category applies to a pathway, the classification does not alter if the initial and final states are interchanged.

Second, ideal pathways offer choices regarding the control variables. Figure 5.13 answers the question raised in Figure 5.3. The ideal pathways are formed either via perfect pathways in tandem or by the most direct route. There are otherwise no limits placed on the X, Y identities; different combinations of p, V, T, S, and so forth are all suitable. Under conditions in which $\Gamma_X = \Gamma_Y$, then $\Omega_X = \Omega_Y$. Note as well that dual-segment pathways offer two choices, upper and lower routes, with identical merit factors.

Chapter 4 discussed how a pathway expresses nonzero mutual information MI_{XY}. It is interesting that exceptions appear in the perfect and worst-case categories. Figure 5.14 shows an isobar with pressure and volume resolution indicated by the dotted lines. In the general case, $MI_{XY \leftrightarrow pV}$ is quantified by:

$$MI_{XY \leftrightarrow pV} = + \sum_{i,j} prob\left(\bar{p}_i + \Delta\bar{p}, \bar{V}_j + \Delta\bar{V}\right) \cdot \log_2 \left[\frac{prob\left(\bar{p}_i + \Delta\bar{p}, \bar{V}_j + \Delta\bar{V}\right)}{prob\left(\bar{p}_i + \Delta\bar{p}\right) \cdot prob\left(\bar{V}_j + \Delta\bar{V}\right)} \right]$$

$$(5.35)$$

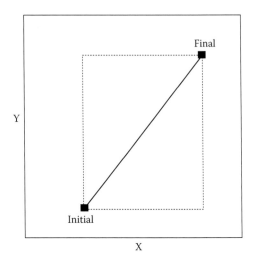

FIGURE 5.13 Answers to the question raised in Figure 5.3. The ideal pathways are formed either via perfect pathways in tandem or by the most direct route. There are no limits placed on the X, Y identities. If $\Gamma_X = \Gamma_Y$, then $\Omega_X = \Omega_Y$. The dual segment pathways offer two choices, upper and lower, with identical merit factors.

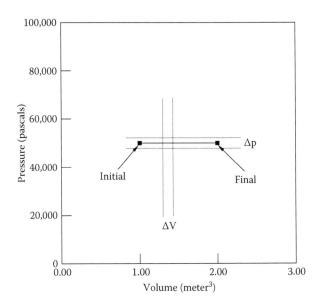

FIGURE 5.14 An isobar with pressure and volume resolution indicated via the dotted lines. $MI_{XY \leftrightarrow pV}$ reduces to zero by Equation (5.35).

where $\bar{p}_i + \Delta\bar{p}$ and $\bar{V}_i + \Delta\bar{V}$ define the state widths and boundaries. But $prob(\bar{p}_i + \Delta\bar{p}) = 1$ across the transformation—the pathway-knowledgeable chemist obtains no information in the statistical sense from measurements using a barometer. By contrast:

$$prob\left(\bar{p}_i + \Delta\bar{p}, \bar{V}_j + \Delta\bar{V}\right) = prob\left(\bar{V}_j + \Delta\bar{V}\right) = \frac{1}{\Gamma_{Y \leftrightarrow V}} \tag{5.36}$$

Thus the logarithm argument in each term of Equation (5.35) reduces to 1, whereby $MI_{XY \leftrightarrow pV} = 0$ is the outcome. This makes sense. If the chemist were to measure the system volume, no pressure information would be obtained as a fringe benefit. By the same token, any measurement of the pressure offers nothing about the volume.

For worst-case pathways, the absence of an algorithm precludes the correlation of state variables. Thus,

$$prob\left(\bar{p}_i + \Delta\bar{p}, \bar{V}_j + \Delta\bar{V}\right) = prob\left(\bar{p}_i + \Delta\bar{p}\right) \times prob\left(\bar{V}_j + \Delta\bar{V}\right) \tag{5.37}$$

which makes the logarithm argument of every term in Equation (5.35) equal to 1 and $MI_{XY \leftrightarrow pV} = 0$. This also agrees with intuition. If the chemist were to measure either p or V, no extra knowledge arrives. How are perfect and worst-case pathways alike, besides sharing state variables X and Y? The answer is that both express $MI_{XY} = 0$.

The mutual information properties of ideal pathways are just as interesting. Refer again to Figure 5.11. There are five p,V segments that lie along A given the resolution indicated. The probability of the chemist observing the system in any one of these is

$$\frac{1}{\Gamma_{X \leftrightarrow p}} = \frac{1}{\Gamma_{Y \leftrightarrow V}} = \frac{1}{5} \tag{5.38}$$

When the system is programmed to travel the straight-line route, the same probability value as Equation (5.38) applies to a p or V measurement made by itself. Thus, for identical resolution of the control variables:

$$prob\left(\bar{p}_i + \Delta\bar{p}, \bar{V}_j + \Delta\bar{V}\right) = \frac{1}{\Gamma_{X \leftrightarrow p}} = prob\left(\bar{V}_j + \Delta\bar{V}\right) = \frac{1}{\Gamma_{Y \leftrightarrow V}} \tag{5.39}$$

with the result that

$$MI_{XY \leftrightarrow pV} = 5 \times \frac{1}{5} \times \log_2\left[\frac{\frac{1}{5}}{\frac{1}{5} \cdot \frac{1}{5}}\right] = \log_2[5] \text{ bits}$$

$$= \log_2[\Gamma_{X \leftrightarrow p}] = \log_2[\Gamma_{Y \leftrightarrow V}] \text{ bits} \tag{5.40}$$

This is important because it shows that for straight-line pathways, the mutual information equates with the Shannon information for each control variable. It answers the question: For what thermodynamic programs are the Shannon and mutual information equal?

Regarding E of Figure 5.11, there are 10 p,V segments at the resolution shown. The probability of observing the system somewhere in any of these is:

$$\frac{1}{2 \cdot \Gamma_{X \leftrightarrow p}} = \frac{1}{2 \cdot \Gamma_{Y \leftrightarrow V}} = \frac{1}{2 \times 5} \tag{5.41}$$

Different probability values, however, apply to any p and V measurements exercised individually. This point was made via Equations (5.26) and (5.27). For the dual-segment case, the mutual information can be shown to have the following form at high resolution (and taking $\Gamma_{X \leftrightarrow p} = \Gamma_{Y \leftrightarrow V}$):

$$MI_{XY \leftrightarrow pV} = \left(\frac{2\Gamma_{X \leftrightarrow p} - 2}{2\Gamma_{X \leftrightarrow p}}\right) \times \log_2 \left[\frac{1}{\frac{1}{2} + \frac{1}{2\Gamma_{X \leftrightarrow p}}}\right] + \frac{1}{\Gamma_{X \leftrightarrow p}} \times \log_2 \left[\frac{\left(\frac{1}{2\Gamma_{X \leftrightarrow p}}\right)}{\left(\frac{1}{2} + \frac{1}{2\Gamma_{X \leftrightarrow p}}\right)^2}\right] \tag{5.42}$$

Then for high-resolution queries, that is, large $\Gamma_{X \leftrightarrow p}$ and $\Gamma_{Y \leftrightarrow V}$, Equation (5.42) reduces approximately to:

$$MI_{XY \leftrightarrow pV} \approx \log_2[2] + \frac{1}{\Gamma_{X \leftrightarrow p}} \times \log_2 \left[\frac{2}{\Gamma_{X \leftrightarrow p}}\right] \approx 1 \text{ bit} \tag{5.43}$$

Equation (5.43) reflects that for ideal pathways of the dual-segment variety, the control variables are correlated— $MI_{XY} > 0$. But the correlations amount to only about 1 bit. Dual-segment pathways are combinations of perfect pathways; the control variables of each are programmed to be as information-strategic and independent as possible.

The major points of this chapter are as follows:

1. Thermodynamic pathways describe programs for taking a system from one state to another. There is always an infinite number of choices, some more attractive than others in their economy. This chapter discussed how to evaluate pathways for both length and information economy using merit factors. The smaller Ω_X, the better the overall economy. $\Omega_X = 0$ offers the maximum economy, whereas $\Omega_X = \infty$ describes the worst case.
2. Economy considerations led to four pathway categories: perfect, worst case, ideal, and nonideal. Every reversible pathway falls into one of these.

3. Pathways express more than one type of information. The perfect, ideal, and worst-case categories offer unusual properties regarding the mutual information.

5.4   SOURCES AND FURTHER READING

Best-practice programming is integral to multiple disciplines. The text by Vajda describes linear programming strategies with applications primarily of the economics variety [2]. Prime factoring figures in modern computer security. The texts by Ribenboim [3], and Crandall and Pomerance [4] are well worth time and attention. A more qualitative book that includes prime factoring issues is by Derbyshire [5].

Regarding convergent syntheses, Fleming describes the challenges and strategies of protein synthesis [6]. The strategies include the use of protecting groups in order to obtain the correct stereochemistry for a protein. Fleming's book is indispensable for its presentation of best-practice syntheses of a variety of compounds across several decades.

The Carnot cycle is treated in all thermodynamic texts. Fermi [7] and Desloge [8] present succinct and illuminating discussions. One should not neglect Carnot's original article available in translation [9]. More contemporary discussions of the Carnot cycle are presented by Finfgeld and Machlup [10], and by Raymond [11] and Van den Broeck [12].

The entropic aspects of heat engines, reversible and otherwise, have been detailed by Berry and coworkers [13].

Random pathways are the trademark of Brownian motion. This subject is treated at length in numerous probability books. Especially recommended are the treatments by Karlin and Taylor [14] and by Resnick [15]. Last, the phase transition behavior described by the van der Waals model is discussed incisively by Stanley [16].

5.5   SUGGESTED EXERCISES

5.1   The prime factors of 159,870,864,030 were listed as 2, 3, 5, 547, 1229, and 7927. How many information bits are obtained upon learning each factor? Please discuss.

5.2   The convergent synthesis of a 256-unit peptide was described. Suppose the chemist was less exacting and prepared a diverse set of peptides in the first stage. For each, the unit number is at least 2 but no more than 6, the number chosen randomly. (a) If the yield of each reaction is 95%, what is the average net yield of the 256-unit peptide? (b) What is the standard deviation in the yield?

5.3   Derive the van der Waals results of Equations (5.4A), (5.4B), and (5.4C).

5.4   Consider the van der Waals isotherm of Figure 5.1 at a pressure resolution window equal to 1% of the total range. (a) What is the value of the merit factor $\Omega_{X \leftrightarrow p}$? (b) How does this result compare to $\Omega_{X \leftrightarrow p}$ for a van der Waals isotherm computed at a temperature above the critical value?

5.5   Consider a system programmed to travel the upper portion of the Figure 5.4 Carnot cycle. Let the initial and final states be i and iii, respectively. (a) What is the value of the merit factor $\Omega_{X \leftrightarrow p}$? Take the pressure resolution to be 1% of the range. (b) What is the value of $MI_{XY \leftrightarrow pV}$ in bits? Take the pressure and volume resolution to be 10% of the respective ranges.

5.6   Direct the previous question to the upper part of the diamond pathway of Figure 5.4. (a) What is the value of the merit factor $\Omega_{X \leftrightarrow p}$? (b) What is the value of $MI_{XY \leftrightarrow pV}$ in bits? Take the resolution window to be the same as in Exercise 5.5.

5.7   Consider an isotherm and adiabat for a monatomic ideal gas. In general, which offers the more favorable $\Omega_{X \leftrightarrow p}$? Please discuss.

5.8   Construct a Brownian path in the pV plane or 1.00 mole of ideal gas. Let the initial state correspond to 10^{-3} meter^3 and 500 K. For each of the 10^4 steps, let two coin tosses—best carried out by computer—decide which variable (p or V) to adjust, along with the direction (positive or negative). Let each relocation of the state point correspond to 0.01% of initial p and V. Plot the pathway and compute $\Omega_{X \leftrightarrow p}$; take the query resolution to be 1% of the total pressure range. (a) If the exercise is carried out multiple times, what average and standard deviation are observed for $\Omega_{X \leftrightarrow p}$? (b) What statistical distribution is formed by the different values of $\Omega_{X \leftrightarrow p}$?

5.9   Consider again pathways A and E of Figure 5.11 for 1.00 mole of monatomic ideal gas. (a) Which expresses the smaller $I_{X \leftrightarrow U}$? (b) Using the initial and final states specified in the figure, construct all the ideal pathways in the TS plane. Which of these expresses the smallest $I_{X \leftrightarrow U}$?

5.10 Derive Equation (5.43) regarding mutual information.

5.11 What is the heat $\rightarrow$ work conversion efficiency of the diamond pathway in Figure 5.4?

REFERENCES

[1] Weast, R. C., ed. 1972. *Handbook of Chemistry and Physics*, p. F61, Chemical Rubber Co., Cleveland, OH.

[2] Vajda, S. 1989. *Mathematical Programming*, Dover, New York.

[3] Ribenboim, P. 1996. *The New Book of Prime Number Records*, Springer, New York.

[4] Crandall, R., Pomerance, C. 2005. *Prime Numbers: A Computational Perspective*, Springer, New York.

[5] Derbyshire, J. 2003. *Prime Obsession*, John Henry Press, Washington, D.C.

[6] Fleming, I. 1973. *Selected Organic Syntheses*, p. 98, Wiley, London.

[7] Fermi, E. 1956. *Thermodynamics*, Dover, New York.

[8] Desloge, E. A. 1968. *Thermal Physics*, Holt, Rinehart, and Winston, New York.

[9] Carnot, S. 1824. *Reflections on the Motive Power of Fire: And Other Papers on the Second Law of Thermodynamics*, Dover, New York.

[10] Finfgeld, C., Machlup, S. 1960. Well-Informed Heat Engine: Efficiency and Maximum Power, *Amer. J. Phys.* 28, 324.

[11] Raymond, R. 1951. The Well-Informed Heat Engine, *Amer. J. Phys.* 19, 109.

[12] Van den Broeck, C. 2007. Carnot Efficiency Revisited, *Adv. Chem. Phys.* 135, 189.

[13] Salamon, P., Nitzan, A., Andresen, B., Berry, R. S. 1980. Minimum Entropy Production and the Optimization of Heat Engines, *Phys. Rev. A* 21, 2115.

[14] Karlin, S., Taylor, H. M. 1975. *A First Course in Stochastic Processes*, 2nd ed., Academic Press, New York.

[15] Resnick, S. L. 1992. *Adventures in Stochastic Processes*, chap. 6, Birkhäuser, Boston.

[16] Stanley, H. E. 1971. *Introduction to Phase Transitions and Critical Phenomena*, Oxford University Press, New York.

6 Thermodynamic Information and Molecules

The preceding chapters focused on thermodynamic state points, both individually and in collections assembled by fluctuations and pathways. The present chapter considers state points at the microscopic level. These points link with the statistical structure presented by molecules and their communication in thermal environments. It is this structure that determines information at the Angstrom scale.

6.1 INFORMATION AT THE MICROSCOPIC SCALE

Chemical thermodynamics concentrates on the states of a system, their description variables, and all matters of work and heat. Information measures the amount of code needed for efficient labeling of the states. The code amounts are important because they tie to the diversity, complexity, and control capacity of the system. These concepts were met qualitatively in Chapter 1, and then quantitatively in Chapter 2 using for example, coins and peptides. Chapter 3 examined the Shannon information associated with fluctuations and equilibrium conditions. Chapters 4 and 5 went on to address $I_{X \leftrightarrow p}$, $KI_{X \leftrightarrow p}$, $MI_{XY \leftrightarrow pV}$, and so forth for reversible pathways programmed for a system. The quantitative examples of the past three chapters have highlighted monatomic gases. SO_2 was the sole polyatomic encountered, and then only briefly via the van der Waals model (cf. Figures 4.10 and 4.17 of Chapter 4). The result is that little attention has been given—not since Chapter 2—to the microscopic level and to molecular structure in general. These, and how they intersect with information, form the themes of the present chapter. To be sure, the microscopic scale is the complicated domain of quantum mechanics and statistical mechanics. The approach taken in this chapter is much simpler as it appeals to idealized models.

To begin, one considers the reaction of Bunsen burners and kitchen stoves:

$$CH_4(g) + 2O_2(g) = CO_2(g) + 2H_2O(g)$$

Standard tables of first-year chemistry texts make contact with the molecular scale by listing standard free energies and enthalpies of formation, and molar entropies:

$CH_4(g)$

$$\Delta G_f^o = -50.7 \text{ kilojoules/mole}$$

$$\Delta H_f^o = -74.8 \text{ kilojoules/mole}$$

$$S^o = 0.1862 \text{ kilojoules/mole} \cdot \text{Kelvin}$$

$O_2(g)$

$$\Delta G_f^o = 0.00 \text{ kilojoules/mole}$$

$$\Delta H_f^o = 0.00 \text{ kilojoules/mole}$$

$$S^o = 0.2050 \text{ kilojoules/mole} \cdot \text{Kelvin}$$

$CO_2(g)$

$$\Delta G_f^o = -394.4 \text{ kilojoules/mole}$$

$$\Delta H_f^o = -393.5 \text{ kilojoules/mole}$$

$$S^o = 0.2136 \text{ kilojoules/mole} \cdot \text{Kelvin}$$

$H_2O(g)$

$$\Delta G_f^o = -228.6 \text{ kilojoules/mole}$$

$$\Delta H_f^o = -241.8 \text{ kilojoules/mole}$$

$$S^o = 0.1887 \text{ kilojoules/mole} \cdot \text{Kelvin}$$

For the reaction, one learns:

$$\Delta G^o = [1 \text{ mole} \cdot (-394.4) + 2 \text{ mole} \cdot (-228.6) - 1 \text{ mole} \cdot (-50.7)$$
$$-2 \text{ mole} \cdot (0.00)] \cdot \frac{\text{kilojoules}}{\text{mole}} = -800.9 \text{ kilojoules} \tag{6.1}$$

$$\Delta H^o = [1 \text{ mole} \cdot (-393.5) + 2 \text{ mole} \cdot (-241.8) - 1 \text{ mole} \cdot (-74.8)$$
$$-2 \text{ mole} \cdot (0.00)] \cdot \frac{\text{kilojoules}}{\text{mole}} = -802.3 \text{ kilojoules} \tag{6.2}$$

$$\Delta S^o = [1 \text{ mole} \cdot (0.2136) + 2 \text{ mole} \cdot (0.1887) - 1 \text{ mole} \cdot (0.1862)$$

$$-2 \text{ mole} \cdot (0.2050)] \cdot \frac{\text{kilojoules}}{\text{mole} \cdot \text{K}} = -0.0052 \frac{\text{kilojoules}}{\text{K}} \qquad (6.3)$$

$$K_p = \exp\left[\frac{-\Delta G^o}{RT}\right] \approx \exp[323.8] \approx 10^{141} \qquad (6.4)$$

These equations are restricted to so-called standard conditions—1.00 atmosphere gases at 298 K. The calculations are nonetheless jumping-off points for approximating real-life energy and material processing. State functions and the properties of equilibrium systems provide far-reaching tools. There will be more to say about the equilibrium constant (K_p) of Equation (6.4) in Chapter 7.

Molecules CH_4, O_2, and so forth are packages of electric charge. Quantum mechanics and statistical mechanics describe them using Hamiltonians, wave functions, and partition functions. At the same time, elementary models count on formula diagrams to portray the Angstrom scale. The gains lie in immediacy and chemical intuition. Hence, the compounds of Bunsen burners and stoves are represented in digital terms:

The benefits include a second approach to the thermodynamics, the first resting on the state functions responsible for Equations (6.1) through (6.4). This is because still more tables of first-year chemistry texts provide the average dissociation energies (D) of atom–bond–atom (ABA) components, for example:

$$D_{C\text{-}H} = 411 \text{ kilojoules/mole}$$
$$D_{C=O} = 803 \text{ kilojoules/mole}$$
$$D_{O\text{-}H} = 464 \text{ kilojoules/mole}$$
$$D_{O=O} = 498 \text{ kilojoules/mole}$$

Using these, the enthalpy of methane combustion can be estimated:

$$\sum_i D_{i,react} - \sum_i D_{i,prod}$$

$$= [4 \text{ mole} \cdot (-411) + 2 \text{ mole} \cdot (-498) - 2 \text{ mole} \cdot (-803) - 4 \text{ mole} \cdot (+464)] \cdot \frac{\text{kilojoules}}{\text{mole}}$$

$$= -822 \text{ kilojoules}$$

$$(6.5)$$

in respectable agreement with Equation (6.2). Note the power and utility of the second approach. In the first, the chemist is restricted by the number of listings. If the standard tables include ΔG_f^o, ΔH_f^o, and S^o for, say, 1-octanol, the chemist can make predictions about the molecule's thermochemistry. If such data are absent, however, the chemist will have to become curious about another molecule. Fortunately, there is plan B.

In traveling the Equation (6.5) route (i.e., plan B), the chemist uses ABA-data to piece together thermochemical properties. There is really no limit on the number of systems the chemist can investigate this way. When electing plan B, the chemist is cognizant of the assumptions and limitations. D_{C-H} is commonly listed as 411 kilo-joules/mole. Yet C-H of methane is not identical to a C-H of propane, 1-octanol, or any other compound. For that matter, there are bond variations of a given type within a compound. For example, C-C adjacent to C=O in cyclohexanone is not identical in length and charge density to C-C opposite C=O. Thus, the dissociation energies of C-C units only a few Angstroms apart are not precisely the same.

In an important way, D tables report averages that are grounded on representative experiments. The numbers vary somewhat from table to table, depending on which data have been compiled. The second approach has currency nonetheless for its simplicity and because of the local nature of most chemical bonds. D tables are invaluable for approximating enthalpy contents, reaction energies, and more, typically within several percent of experimental values. As discussed in Chapter 1, diagrams composed of ABA units are really without peer in the ability to capture the Angstrom scale in digital terms.

Regardless of approach, the chemist appreciates the importance of the molecular scale. Combustion of 1.00 mole of methane yields approximately 800 kilojoules of free energy. This should be compared to the work afforded by isothermal expansion of 1.00 mole of ideal gas at 298 K:

$$W_{perf} = +nRT \cdot \log_e \left(\frac{V_{final}}{V_{initial}} \right)$$

$$\approx 1.00 \text{ mole} \times 8.31 \frac{\text{joules}}{\text{mole} \cdot K} \times 298 \text{ } K \times \log_e \left(\frac{V_{final}}{V_{initial}} \right) \qquad (6.6)$$

$$\approx 2476 \text{ joules} \times \log_e \left(\frac{V_{final}}{V_{initial}} \right)$$

Even with factor of 10 volume increases, the energy is less than 1% of that offered by combustion of 1.00 mole of CH_4. Clearly, molecules and their reactions are stellar resources when it comes to work and heat. Further, standard tables of ΔG_f^o, ΔH_f^o, and S^o (plan A) and bond dissociation energies (plan B) provide state descriptors that burrow deeper than p, V, and other macroscopic functions.

Plan B is without limit in the systems that can be addressed. At the same time, it has a drawback, which is perhaps subtle but understood at once by example. The structure of n-butane is represented by:

Tables report $D_{C\text{-}C} = 346$ kilojoules/mole along with $D_{C\text{-}H} = 411$ kilojoules/mole. Thus the sum (D_{total}), average ($<D>$), and standard deviation (σ_D) of the bond energies are:

$$D_{total} = [3 \text{ mole} \cdot (346) + 10 \text{ mole} \cdot (411)] \cdot \frac{\text{kilojoules}}{\text{mole}} = 5148 \text{ kilojoules} \qquad (6.7)$$

$$<D> = \frac{[3 \text{ mole} \cdot (346) + 10 \text{ mole} \cdot (411)] \cdot \dfrac{\text{kilojoules}}{\text{mole}}}{13} = 396 \text{ kilojoules} \qquad (6.8)$$

$$\sigma_D = \sqrt{\frac{\sum_i (D_i - 396)^2}{13 - 1}} \text{ kilojoules} = 28.5 \text{ kilojoules} \qquad (6.9)$$

Equations (6.7) through (6.9) thereby connect with a sample of n-butane at the molecular level but not uniquely. Via the ABA energies, one arrives at identical values of D_{total}, $<D>$, and σ_D for iso-butane having the formula diagram:

Many more examples can be constructed. To cite an extreme case, there are over 4 million isomers allowed by the formula $C_{30}H_{62}$. D_{total}, $<D>$, and σ_D according to plan B are identical for all. This is unfortunate, and the fallout is not restricted to alkanes. On the one hand, plan B is without limit in providing descriptors for the molecular scale. On the other hand, it does not afford the same discrimination as plan A.

This brings us to the central idea of Chapter 6, namely, the use of ABA units C-C, C-H, and so forth, to describe states of a system at the microscopic level, in a way that always distinguishes molecules by their electronic structure. The diagram for any of the four million versions of $C_{30}H_{62}$ offers the chemist unique facts and data information, for example:

When information of the statistical type is incorporated, the diagram locates state points for the molecule that are also unique. The equivalent statement holds for n-butane, iso-butane, 1-octanol, and so forth—any compound of interest to the chemist.

The point locations arrive by an elementary view of molecular communication in thermal environments.

It should be emphasized, however, that a system's microscopic states are not as straightforward as they might appear, even in the first-year arena of formula diagrams and reactant → product statements. This is true for compounds as small as methane and everyday processes such as combustion. The reason is that virtually all microscopic events are grounded upon the electronic messages carried by molecules, and transmitted and registered via collisions. Thermal energy stored in molecular translations and rotations provides the carrying, transmission, and reception power. In a system as familiar as a Bunsen burner flame, there are no fewer than 14 message-bearing components: N_2, O_2, CO_2, Ar, Ne, Kr, Xe, H_2, CH_4, He_2, H_2O, ethane, propane, and ethanethiol [1]. The reaction that liberates heat and enriches labs with CO_2, steam, and other products happens on account of binary collisions. Yet, with $\kappa = 14$ vehicles in play, there are quite a few two-party combinations, namely,

$$\sum_{j=0}^{\kappa-1} (\kappa - j) = 14 + 13 + \cdots + 1 = 105 \qquad (6.10)$$

Each combination provides its own brand of electronic data processing: N_2-O_2, N_2-CO_2, N_2-Ar, and so on. The message sending and receiving are frequent given that the collision rate for each molecule ranges between 10^9 and 10^{11} sec^{-1}.

Consider a communication event portrayed in Figure 6.1. Shown are a generic interaction potential $\Phi(r)$, and CH_4 and N_2 (as in a Bunsen burner flame) at separation distance (r), poised for collision. At large r, the potential energy of interaction is effectively zero as marked by the dotted horizontal line. As the molecules approach each other, however, their electronic energy decreases due to forces of mutual attraction. As the encounter proceeds, energy is redistributed in the rotational and translational, and, to a minor extent, vibrational degrees of freedom. New sites of both parties experience contact—attraction and repulsion—as part of a dynamic complex. Since the union marks a local reduction of the system volume, there is a decrease of entropy. The entropy diminution brings with it a trapping of information that is unique to CH_4 and N_2.

Under most conditions, chemical reactions occur rarely compared with thermal collisions. This owes to the stability of molecules and their activation requirements—a bond in CH_4 must be fractured in order to become amenable to oxidation or other electronic conversion. Yet, given collision frequencies such as in a flame, momentary fluctuations of the entropy (among other thermodynamic quantities) are the predominant events. The irreversibilities of combustion—heat, CO_2, and steam production—are scarce by comparison. For the majority of encounters, the charge packages end up compressing one another slightly. This causes the interaction energy (as reflected by $\Phi(r)$) to increase dramatically above the null values applicable at large r. Quickly the molecules separate, redistributing energy yet again in the translational and rotational motion. Whatever information had been trapped via the entropy reduction is lost in subsequent collisions. When it comes to the microscopic level, memories within a system tend to be short-lived.

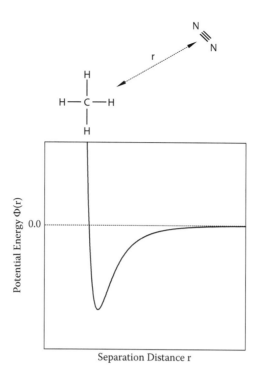

FIGURE 6.1 Elements of communication among molecules. Shown are CH_4 and N_2 poised for collision. The generic potential $\Phi(r)$ describes the interaction of the molecules.

There is another point of Figure 6.1. CH_4 does not transfer an electronic message to N_2 or any other molecule. Further, there is no recognition extended by N_2 or other compound toward CH_4. Rather, the colliding parties and their neighbors act collectively as message sources, receivers, and broadcast channels. At the macroscopic level, the borders that separate the components of communication are distinct; a gas transmits temperature data to a thermometer, pressure data to a barometer, and so forth. The Angstrom scale is exceptional because the components are highly integrated.

The molecular scale is further complicated in that a single $\Phi(r)$ cannot tell the whole story. Since CH_4 and N_2 make contact in a flame at different speeds, rotations, and trajectories, there are inevitable variations of the Figure 6.1 potential. In particular, the placement and steepness of the vertical portion of $\Phi(r)$ are fluid as shown in Figure 6.2. The identical statement holds for the well depth. The result is that the number of possible messages considerably exceeds the binary combinations noted in Equation (6.10). Collisions between CH_4 and N_2 do not offer a single message, but rather an extended set. There is a set of electronic messages for every combination of a system, irrespective of the phase nature. This statement is equally valid for higher-order collisions (e.g., ternary), although these will not be considered due to their rarity compared with binary.

Information is a featured commodity of all molecules. Compounds large and small carry, communicate, and process it. The hypothetical cell of Chapter 2 was

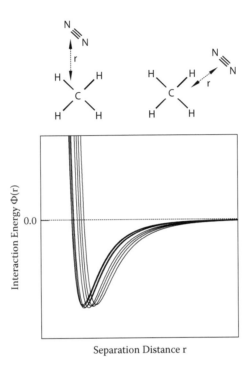

FIGURE 6.2   Molecular communication revisited. Colliding CH_4 and N_2 demonstrate multiple trajectories and interaction potentials.

only capable of synthesizing peptides because it hosted certain polynucleotides. The peptide with formula:

VVRRVRVRRVVVRRRVRVVRRRVVRVRRRRVRRRVRVVRVVVVVVVVRRRRVR-
RRVRVVVVRRVRRVVRVVVVVVRRVVVVVVVVVVRVVVVVVVRVVVVVR

could have originated by the cell's processing of the polynucleotide:

$^{5'}$GUAGUGCGUCGGGUGCGUGUCCGCCGCGUAGUUGUCCGCCGUCGGGUACGAGUU
GUCCGCCGCCGCGUCGUUCGCGUCCGGCGCCGACGGGUUCGCCGGCGAGUACGCGU
AGUCCGCGUUGUAGUUGUAGUUGUCGUCGUGCGUCGACGCCGUGUGCGACGACGAG
UACGUGUCGUGGUCGUUCGCCGGGUCCGACGUGUCGUUCGGGUUGUAGUAGUCGUA
CGGCGGGUCGUUGUAGUAGUCGUCGUCGUAGUCGUACGUGUCGUAGUCGUGGUGGU-
CCGAGUUGUCGUAGUAGUACGG$^{3'}$

or perhaps

$^{5'}$GUCCGUCGCCGUGUGGUCGUUGUCCGGGUGGUACGUGUGGUCCGCGUUCGU-
GUGCGACGGGUAGUGCGUCGCGUCGUCGUUCGUGUUGUCGUUCGUCGAGUUGUGCG
ACGCCGUGUUCGCGUGGUGGUGCGUGUCGUGGUUCGAGUCCGGCGUGUU-
GUUCGCGUUGUAGUUCGUGUUCGCGUCCGCCGCGUCGUUCGACGUCGCCGGGUC-

CGAGUGCGGCGUGUCGUCCGUGUUCGGGUCCGCCGGCGAGUCCGGCGCGUUCGGC-
GAGUGCGUGUCGUGGUGCGCCGACGCGUUGUACGG$^{3\prime}$

There are a large number of possibilities given that the genetic code allots four codons each for incorporating R and V into a protein [2]:

R: CGU, CGC, CGA, CGG
V: GUU, GUC, GUA, GUG

Thus, a 100-unit peptide restricted to V and R offers $2^{100} \approx 10^{30}$ possible information-bearing precursors. The number of precursors of the polynucleotides is astronomically greater: $8^{100} \approx 2 \times 10^{90}$. Here A, G, U, and C are the respective abbreviations for the base units of messenger RNA: adenine, guanine, uracil, and cytosine. The point of this digression is that the messages of one molecule generally descend from the information of others; in the cell's case, proteins and polynucleotides. Further, information processing need not involve molecules of different families. The origins of the aforementioned poly(ribo)nucleotides are themselves poly(deoxyribo)nucleotides:

$^{3\prime}$CATGCTGCCGCTCAGCACCATCAAGCGCATCATGCACAGCATGCGCAGGCG-
CACGCGGCCCAACACGCTGCCCATCATCAAGCACATCAGCAGGCTGCTCACCACGC
TGCTGCTCAGGCGCAACATCAGGCTCAGCACCAAGCGCAGGCTGCCCAGCATGCG-
CAGCAACAGGCCCAGGCACACGCGGCTCAGCATGCAGCCGCGGCCCAAGCG-
CAAGCCGCTCAGCACGCGCAAGCACATGCGGCCGCGCACGCGGCTCATGCTGC-
CCATGCCCACCATCAAGCCGCTGCGCACCACGCA$^{5\prime}$

and

$^{3\prime}$CAGGCAGCGGCACACCAGCAACAGGCCCACCATGCACACCAGGCGCAAGCA-
CACGCTGCCCATCACGCAGCGCAGCAGCAAGCACAACAGCAAGCAGCTCAACACGC
TGCGGCACAAGCGCACCACCACGCACAGCACCAAGCTCAGGCCGCACAACAAGCG-
CAACATCAAGCACAAGCGCAGGCGGCGCAGCAAGCTGCAGCGGCCCAGGCT-
CACGCCGCACAGCAGGCACAAGCCCAGGCGGCCGCTCAGGCCGCGCAAGCCGCT-
CACGCACAGCACCACGCGGCTGCGCAACATGCC$^{5\prime}$

Such are the products of syntheses catalyzed by polymerase enzymes where T is the abbreviation for thymine base units. Along similar lines, there are protein–protein interactions critical to enzyme regulation—information impacts information. It is the field of bioinformatics that concentrates on polynucleotides and proteins, their structure, function, and host organisms. The actions of these large compounds encompass the full range of message writing, copying, editing, and deleting. One refers to the end-of-chapter references for comprehensive treatments.

Concerning molecules more modest in size, the chemical fields are no less active. This is because electronic states and information combine to make a wide-angle lens. The research has ranged from graph topology analysis to the information of density functionals. The advances have been motivated by the need to understand energy dispersal among molecules, structure-activity relations, and data mining for pharmaceutical applications. Information of the statistical type sheds light where there are states to enumerate and correlate.

To keep the focus on thermodynamics, one looks again at the ABA nature of molecules—assemblies of C-C, C-H, and so on. One further keeps in mind the lessons of Figures 6.1 and 6.2, namely, that molecular communication occurs by thermal collisions.

The following are elementary reaction statements:

For each case, the reactants convert to products because of free energy losses. Yet insofar as information is concerned, the critical events include the message transmission and reception that precede the losses. Reactions are rare compared with electronic information processing. In a sample of Br_2 and 1-butene, the halogen reacts at the site of unsaturation. In a solution of Cl_2 and cyclopentanone and its enol tautomer, a halogen is substituted for a hydrogen atom at the site alpha to the carbon-oxygen bond. These statements describe the *ultimate* actions of the reagents. Yet *all* collision sites of the molecules are sampled in thermal environments regardless of chemical activity. Chemical reactions happen because they increase the total entropy. Their selectivity, however, arises from the information expressed by the participant molecules. The message processing is as complicated as indicated in Figure 6.2. Fortunately for the chemist, formula diagrams provide digital reductions of facts and data. Further consideration of communication mechanisms leads to information in the statistical sense.

C. H. Bennett examined molecular information processing at length in the *Thermodynamics of Computation* [3]. His discussion looked to biopolymers DNA and messenger RNA for showcase examples. Importantly, thermal energy governs the state sampling. The result is that electronic messages are communicated by random motion and collisions—erratic, bumpy random walks of the molecules. The walks ensure that all the possible contact sites are sampled thoroughly in a heat-filled environment. Bennett characterized the information processing as *Brownian computation*. Molecules operate as Brownian computers because their state transitions occur quite by thermodynamic accidents—local fluctuations enabled by translations, rotations, and collisions.

As discussed in Chapter 2, information is the by-product of states, mechanisms for communication, and uncertainty. Chapters 3 through 5 directed these concepts to the macroscopic states of elementary systems. At the Angstrom level, one looks to the collisions afforded by molecules—electronic contacts of ABA sites C-C, C-H, C-O, and so forth—for states and mechanisms. It is the random nature that supplies the uncertainty.

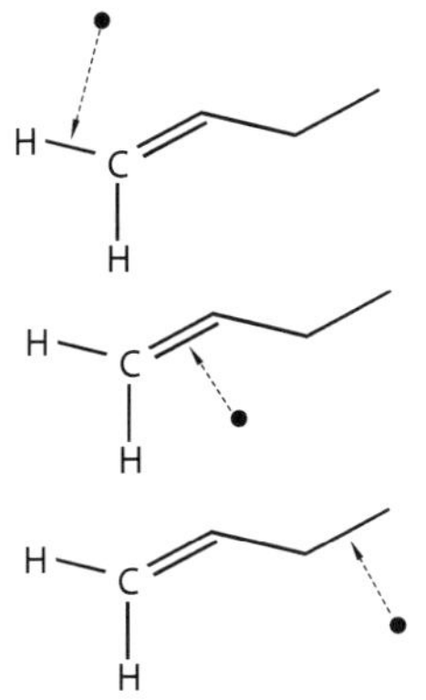

FIGURE 6.3 Transmission and registration of electronic messages. 1-butene poses three types of message-bearing sites: C-H, C=C, and C-C. An electronic message is transmitted and registered upon collision with another party such as a helium atom represented by the filled circle.

The communication and registration of an ABA unit is imagined as a low-energy thermal contact between the molecule of interest and an atom such as helium. The event involves energy being redistributed and new sites becoming available for immediate contact; a molecule does not suffer merely one collision at one discrete site but rather an extended sequence. In keeping with idealizations of the macroscopic level, all collision sites, and thus electronic messages, are viewed as equally accessible. However, the probability of transmitting and registering a C-H, C-C, C=C, and so on is taken as proportional to its occurrence in a random (i.e., thermal) walk over the molecule. Figure 6.3 illustrates the essentials of the model. There are innumerable messages afforded by a compound such as 1-butene. In the model, however, the units that compose the messages are pooled into three distinctive sets having digital labels C-H, C-C, and C=C. Along the same lines, there are three distinct message units carried by cyclopentanone with labels C-H, C-C, and C=O. It is easy to count sets and identify labels for a molecule by inspecting the formula diagram.

At the macroscopic level, a system's nearest-neighbor states are accessed by fluctuations and structured programs such as isotherms. There is a parallel at the Angstrom scale. If one ABA site is registered by a thermal collision, the thermally most probable site to be acted upon, and to affect the electronic message, subsequently will be a nearest neighbor. Since there is no reason to hold one site more significant than another, all nearest neighbors should be viewed as equally probable for thermal contact. This idea is illustrated in Figure 6.4. If the atom makes contact with C-H, C=C, or C-C as in the upper, middle, and lower diagrams, respectively, the most probable ensuing collision (interaction plus message transmission and registration) will involve nearest ABAs determined by the molecular structure.

Polynucleotides accommodate four different nucleobase sites or message units. They offer enormous diversity by extended sequences, for example, … GUAGUGCGUCGG…. In a parallel way, molecules offer a richness of ABA-contact sequences … (C-C)(C-H)(C-H)(C-C)… and thus electronic messages. As

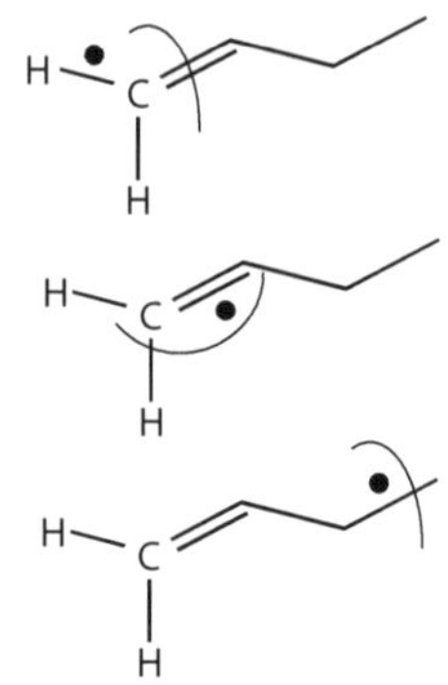

FIGURE 6.4 Sequential transmission and registration of messages. A thermal collision can occur at a site such as C-H in the uppermost diagram. Translation and rotation (indicated by the arc symbols) of the molecule will then engage the message of the neighbor site such as C=C or C-H. If the collision initiates at C=C as in the middle diagram, the succeeding event involves C-H or C-C. If the collision occurs at C-C as in the lower-most diagram, the succeeding event involves C-H or C-C.

with nucleobase sequences, the messages of a molecule can be specified using the digital units that encode the formula diagram. A given sequence describes an electronic message that is truly unique to the source, say, 1-butene as opposed to cyclopentanone. The collection of possible sequences details the electronic message space for the molecule. As with polynucleotides, there is overlap of the space for one molecule to the next. This is because different molecules demonstrate ABAs in common, for example, 1-butene, cyclopentanone, and ethane all host C-C and C-H units. The uncommon units, however, along with their thermal walk particulars confer uniqueness. As shown in the next section, it is straightforward to establish the message space—the collection of possible messages—of everyday molecules and to quantify the Shannon and mutual information. Polynucleotides are hard-copy message tapes processed via polymerases and other cellular components. For Brownian computers like 1-butene, the message tapes must be constructed and analyzed in virtual terms.

6.2   MOLECULAR MESSAGE TAPES: TECHNICAL CONSIDERATIONS

At the macroscopic level, information is quantified via multivariable functions, line integrals, and probability distributions—the tools central to Chapters 3 through 5. The thermodynamic situations that admit analytical solutions are rare. Approximations are necessary with the assistance of computer programming and spreadsheets.

At the molecular scale, matters are different in certain respects while similar in others. There are no multivariable functions or integrals to worry about. Yet pencil-and-paper calculations range from the impractical to impossible. Examining a molecule as a Brownian computer requires construction of a random walk over the finite collection of ABA sites. A few essentials can be grasped via Figures 6.3 and 6.4. To go further, one considers again a Bunsen burner flame that hosts ethane, propane, and ethanethiol (among other compounds) in small amounts. Fragments of random walks over these molecules can be represented in diagram terms as

Ethane:

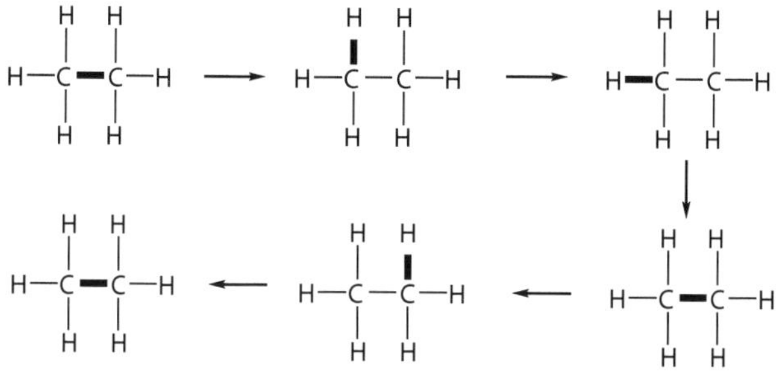

Propane:

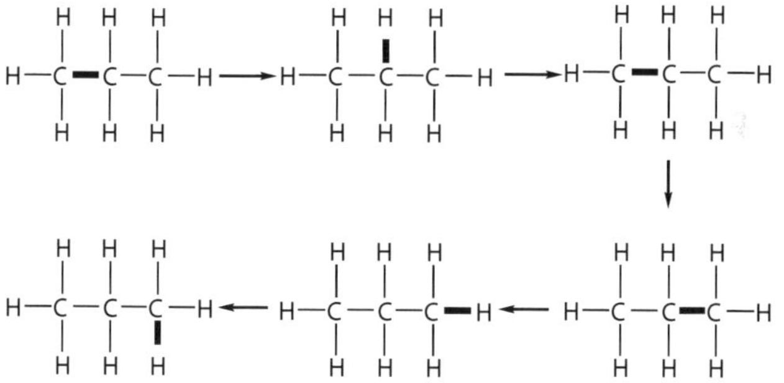

Ethanethiol:

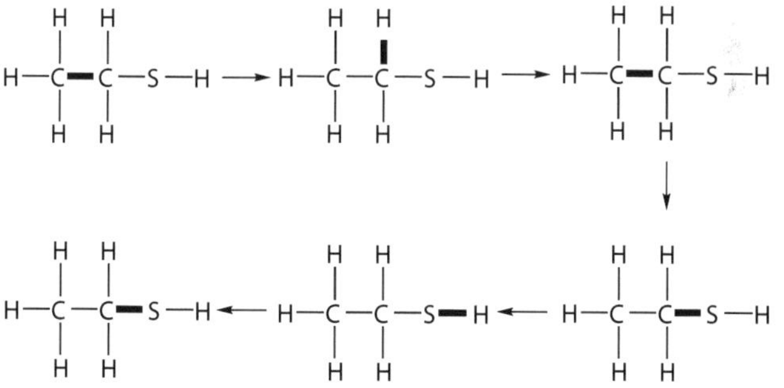

The electronic site in contact with a hypothetical colliding element (not shown) has been indicated using boldface. Each step accesses a nearest-neighbor ABA of the formula diagram representing the molecule. In real Bunsen burners, the contacts are made by thermal collisions. Neither the motion (translational and rotational) nor interaction specifics are contained in the model. As a consequence, all the ABA sites are treated in equal-likelihood terms, and their nature is unaltered during transmission and registration. This means that C-C never becomes more significant than C-H or other units; C-C never starts to look like C-S and vice versa. Information in the statistical sense requires systems whose states are robust and accommodating of digital labels.

Diagrams capture the thermal walks pictorially. Quantitative representations are obtained with the help of atom vectors and bond matrices. For an example, let each atomic symbol in the diagram for ethane be indexed as follows:

$$
\begin{array}{ccc}
^3\text{H} & \text{H}^6 \\
| & | \\
^4\text{H}-\text{C}^1-\text{C}^2-\text{H}^7 \\
| & | \\
_5\text{H} & \text{H}_8
\end{array}
$$

The corresponding atom vector and bond matrix become

$$
\begin{bmatrix} 6 \\ 6 \\ 1 \\ 1 \\ 1 \\ 1 \\ 1 \\ 1 \end{bmatrix}
\begin{bmatrix}
0 & 1 & 1 & 1 & 1 & 0 & 0 & 0 \\
1 & 0 & 0 & 0 & 0 & 1 & 1 & 1 \\
1 & 0 & 0 & 0 & 0 & 0 & 0 & 0 \\
1 & 0 & 0 & 0 & 0 & 0 & 0 & 0 \\
1 & 0 & 0 & 0 & 0 & 0 & 0 & 0 \\
0 & 1 & 0 & 0 & 0 & 0 & 0 & 0 \\
0 & 1 & 0 & 0 & 0 & 0 & 0 & 0 \\
0 & 1 & 0 & 0 & 0 & 0 & 0 & 0
\end{bmatrix}
$$

The vector components are the atomic numbers 6, 6, 1, ..., 1 corresponding to letters of the formula diagram. The matrix, in turn, specifies all the linkages and covalent bond orders. Since ethane is composed of eight atoms, the vector hosts eight entries while the matrix dimensions are 8×8. There are notable characteristics of the matrix. Clearly, the majority of components are zero, thus rendering a certain sparseness. The diagonal elements are mandatorily zero since an atom cannot bind to itself. The matrix is square symmetric with a determinant value of zero.

The idealizations are apparent. All the atoms of a molecule interact electronically with one another. But as with the formula diagram, the matrix portrays the covalent interactions to be overriding. The diagram and matrix model the Angstrom-scale electronics as short range and local in attraction and repulsion.

It is straightforward to construct atom vectors and bond matrices based on formula diagrams. Additional examples are given as follows.

Acetone:

$$
\begin{array}{ccccc}
_5\text{H} & & \text{O}\,^1 & & \text{H}\,^8 \\
| & & \| & & | \\
^6\text{H}-\text{C}_3 & - & \text{C}_2 & - & \text{C}_4-\text{H}_9 \\
| & & & & | \\
\text{H} & & & & \text{H} \\
_7 & & & & _{10}
\end{array}
$$

$$
\begin{bmatrix} 8 \\ 6 \\ 6 \\ 6 \\ 1 \\ 1 \\ 1 \\ 1 \\ 1 \\ 1 \end{bmatrix}
\begin{bmatrix}
0 & 2 & 0 & 0 & 0 & 0 & 0 & 0 & 0 & 0 \\
2 & 0 & 1 & 1 & 0 & 0 & 0 & 0 & 0 & 0 \\
0 & 1 & 0 & 0 & 1 & 1 & 1 & 0 & 0 & 0 \\
0 & 1 & 0 & 0 & 0 & 0 & 0 & 1 & 1 & 1 \\
0 & 0 & 1 & 0 & 0 & 0 & 0 & 0 & 0 & 0 \\
0 & 0 & 1 & 0 & 0 & 0 & 0 & 0 & 0 & 0 \\
0 & 0 & 1 & 0 & 0 & 0 & 0 & 0 & 0 & 0 \\
0 & 0 & 0 & 1 & 0 & 0 & 0 & 0 & 0 & 0 \\
0 & 0 & 0 & 1 & 0 & 0 & 0 & 0 & 0 & 0 \\
0 & 0 & 0 & 1 & 0 & 0 & 0 & 0 & 0 & 0
\end{bmatrix}
$$

Acetic acid:

$$
\begin{bmatrix} 8 \\ 8 \\ 6 \\ 6 \\ 1 \\ 1 \\ 1 \\ 1 \end{bmatrix}
\begin{bmatrix}
0 & 0 & 2 & 0 & 0 & 0 & 0 & 0 \\
0 & 0 & 1 & 0 & 1 & 0 & 0 & 0 \\
2 & 1 & 0 & 1 & 0 & 0 & 0 & 0 \\
0 & 0 & 1 & 0 & 0 & 1 & 1 & 1 \\
0 & 1 & 0 & 0 & 0 & 0 & 0 & 0 \\
0 & 0 & 0 & 1 & 0 & 0 & 0 & 0 \\
0 & 0 & 0 & 1 & 0 & 0 & 0 & 0 \\
0 & 0 & 0 & 1 & 0 & 0 & 0 & 0
\end{bmatrix}
$$

Ethanethiol:

$$
\begin{bmatrix} 16 \\ 6 \\ 6 \\ 1 \\ 1 \\ 1 \\ 1 \\ 1 \\ 1 \end{bmatrix}
\begin{bmatrix}
0 & 1 & 0 & 1 & 0 & 0 & 0 & 0 & 0 \\
1 & 0 & 1 & 0 & 1 & 1 & 0 & 0 & 0 \\
0 & 1 & 0 & 0 & 0 & 0 & 1 & 1 & 1 \\
1 & 0 & 0 & 0 & 0 & 0 & 0 & 0 & 0 \\
0 & 1 & 0 & 0 & 0 & 0 & 0 & 0 & 0 \\
0 & 1 & 0 & 0 & 0 & 0 & 0 & 0 & 0 \\
0 & 0 & 1 & 0 & 0 & 0 & 0 & 0 & 0 \\
0 & 0 & 1 & 0 & 0 & 0 & 0 & 0 & 0 \\
0 & 0 & 1 & 0 & 0 & 0 & 0 & 0 & 0
\end{bmatrix}
$$

The indexing of the atom symbols is arbitrary, although it is convenient to pair the major atoms (i.e., all but H) with the uppermost vector and matrix slots. One can then apply a computer subroutine to fill the hydrogen slots based on chemical valence rules.

To initiate a random walk—the transmission and registration of electronic messages by thermal collision—one chooses an atom, say, 2, for ethane via a random number generator (cf. Chapter 2). The corresponding entry is tagged in the vector—designated by a wavy line in the following; the nonzero entries in the same row of the matrix are then counted and labeled. The results of these operations can be represented as

$$
\begin{bmatrix} 6 \\ \tilde{6} \\ 1 \\ 1 \\ 1 \\ 1 \\ 1 \\ 1 \end{bmatrix}
\begin{bmatrix}
0 & 1 & 1 & 1 & 1 & 0 & 0 & 0 \\
1^1 & 0 & 0 & 0 & 0 & 1^2 & 1^3 & 1^4 \\
1 & 0 & 0 & 0 & 0 & 0 & 0 & 0 \\
1 & 0 & 0 & 0 & 0 & 0 & 0 & 0 \\
1 & 0 & 0 & 0 & 0 & 0 & 0 & 0 \\
0 & 1 & 0 & 0 & 0 & 0 & 0 & 0 \\
0 & 1 & 0 & 0 & 0 & 0 & 0 & 0 \\
0 & 1 & 0 & 0 & 0 & 0 & 0 & 0
\end{bmatrix}
$$

The random number generator is then used to select one of the labeled entries of the second row of the matrix, say, 3 corresponding to H unit 7. In so doing, an ABA site of ethane has been established and needs to be labeled as such. In vector, and matrix terms, one has

$$
\begin{bmatrix} 6 \\ 6 \\ 1 \\ 1 \\ 1 \\ 1 \\ \tilde{1} \\ 1 \end{bmatrix}
\begin{bmatrix}
0 & 1 & 1 & 1 & 1 & 0 & \tilde{0} & 0 \\
1 & 0 & 0 & 0 & 0 & 1 & \tilde{1} & 1 \\
1 & 0 & 0 & 0 & 0 & 0 & 0 & 0 \\
1 & 0 & 0 & 0 & 0 & 0 & 0 & 0 \\
1 & 0 & 0 & 0 & 0 & 0 & 0 & 0 \\
0 & \tilde{1} & 0 & 0 & 0 & 0 & 0 & 0 \\
0 & \tilde{1} & 0 & 0 & 0 & 0 & 0 & 0 \\
0 & 1 & 0 & 0 & 0 & 0 & 0 & 0
\end{bmatrix}
$$

The initial entry logged on the message tape becomes C-H.

The next step selects a nearest-neighbor site at random. The present site of the walker must be indicated by another tag, represented with an overhead curve in the following. The eligible jump sites are then counted and labeled. In vector and matrix terms, one has

$$
\begin{bmatrix} 6 \\ \hat{6} \\ 1 \\ 1 \\ 1 \\ 1 \\ \hat{1} \\ 1 \end{bmatrix}
\begin{bmatrix}
0 & 1 & 1 & 1 & 1 & 0 & \hat{0} & 0 \\
1^1 & 0 & 0 & 0 & 0 & 1^2 & \hat{1} & 1^3 \\
1 & 0 & 0 & 0 & 0 & 0 & 0 & 0 \\
1 & 0 & 0 & 0 & 0 & 0 & 0 & 0 \\
1 & 0 & 0 & 0 & 0 & 0 & 0 & 0 \\
0 & 1 & 0 & 0 & 0 & 0 & 0 & 0 \\
0 & \hat{1} & 0 & 0 & 0 & 0 & 0 & 0 \\
0 & 1 & 0 & 0 & 0 & 0 & 0 & 0
\end{bmatrix}
$$

A random number generator is then used to choose one of the superscript-labeled entries, say, 1 corresponding to carbon unit 2. In so doing, a follow-up site of the walker is established. In vector and matrix terms, one has

$$
\begin{bmatrix} \tilde{6} \\ \tilde{6} \\ 1 \\ 1 \\ 1 \\ 1 \\ 1 \\ 1 \end{bmatrix}
\begin{bmatrix}
0 & \tilde{1} & 1 & 1 & 1 & 0 & 0 & 0 \\
\tilde{1} & 0 & 0 & 0 & 0 & 1 & 1 & 1 \\
1 & 0 & 0 & 0 & 0 & 0 & 0 & 0 \\
1 & 0 & 0 & 0 & 0 & 0 & 0 & 0 \\
1 & 0 & 0 & 0 & 0 & 0 & 0 & 0 \\
0 & 1 & 0 & 0 & 0 & 0 & 0 & 0 \\
0 & 1 & 0 & 0 & 0 & 0 & 0 & 0 \\
0 & 1 & 0 & 0 & 0 & 0 & 0 & 0
\end{bmatrix}
$$

The message unit logged by the message tape becomes C-C. It should be apparent that repeated tagging, counting, and random selection results in a Brownian walk over the molecule. The tape tracks the sequence of collisions via nearest-neighbor ABA-message units. Random walks were illustrated earlier for ethane, propane, and ethanethiol in formula diagram terms. When vectors and matrices are used to encode the walks, the results appear as

Ethane:

$$
\begin{bmatrix} \tilde{6} \\ \tilde{6} \\ 1 \\ 1 \\ 1 \\ 1 \\ 1 \\ 1 \end{bmatrix}
\begin{bmatrix}
0 & \tilde{1} & 1 & 1 & 1 & 0 & 0 & 0 \\
\tilde{1} & 0 & 0 & 0 & 0 & 1 & 1 & 1 \\
1 & 0 & 0 & 0 & 0 & 0 & 0 & 0 \\
1 & 0 & 0 & 0 & 0 & 0 & 0 & 0 \\
1 & 0 & 0 & 0 & 0 & 0 & 0 & 0 \\
0 & 1 & 0 & 0 & 0 & 0 & 0 & 0 \\
0 & 1 & 0 & 0 & 0 & 0 & 0 & 0 \\
0 & 1 & 0 & 0 & 0 & 0 & 0 & 0
\end{bmatrix}
\rightarrow
\begin{bmatrix} 6 \\ \tilde{6} \\ 1 \\ 1 \\ 1 \\ \tilde{1} \\ 1 \\ 1 \end{bmatrix}
\begin{bmatrix}
0 & 1 & 1 & 1 & 1 & 0 & 0 & 0 \\
1 & 0 & 0 & 0 & 0 & \tilde{1} & 1 & 1 \\
1 & 0 & 0 & 0 & 0 & 0 & 0 & 0 \\
1 & 0 & 0 & 0 & 0 & 0 & 0 & 0 \\
1 & 0 & 0 & 0 & 0 & 0 & 0 & 0 \\
0 & \tilde{1} & 0 & 0 & 0 & 0 & 0 & 0 \\
0 & 1 & 0 & 0 & 0 & 0 & 0 & 0 \\
0 & 1 & 0 & 0 & 0 & 0 & 0 & 0
\end{bmatrix}
\rightarrow
\begin{bmatrix} 6 \\ \tilde{6} \\ 1 \\ 1 \\ 1 \\ 1 \\ \tilde{1} \\ 1 \end{bmatrix}
\begin{bmatrix}
0 & 1 & 1 & 1 & 1 & 0 & 0 & 0 \\
1 & 0 & 0 & 0 & 0 & 1 & \tilde{1} & 1 \\
1 & 0 & 0 & 0 & 0 & 0 & 0 & 0 \\
1 & 0 & 0 & 0 & 0 & 0 & 0 & 0 \\
1 & 0 & 0 & 0 & 0 & 0 & 0 & 0 \\
0 & 1 & 0 & 0 & 0 & 0 & 0 & 0 \\
0 & \tilde{1} & 0 & 0 & 0 & 0 & 0 & 0 \\
0 & 1 & 0 & 0 & 0 & 0 & 0 & 0
\end{bmatrix}
$$

$$\text{C-C} \qquad \rightarrow \qquad \text{C-H} \qquad \rightarrow \qquad \text{C-H}$$

Propane:

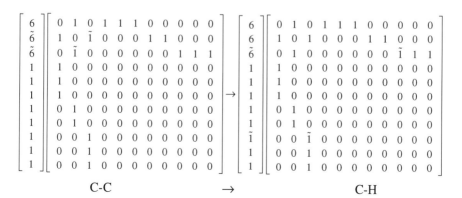

$$\text{C-C} \qquad \rightarrow \qquad \text{C-H}$$

Ethanethiol:

$$
\begin{bmatrix} 16 \\ \tilde{6} \\ \tilde{6} \\ 1 \\ 1 \\ 1 \\ 1 \\ 1 \\ 1 \end{bmatrix}
\begin{bmatrix}
0 & 1 & 0 & 1 & 0 & 0 & 0 & 0 & 0 \\
1 & 0 & \tilde{1} & 0 & 1 & 1 & 0 & 0 & 0 \\
0 & \tilde{1} & 0 & 0 & 0 & 0 & 1 & 1 & 1 \\
1 & 0 & 0 & 0 & 0 & 0 & 0 & 0 & 0 \\
0 & 1 & 0 & 0 & 0 & 0 & 0 & 0 & 0 \\
0 & 1 & 0 & 0 & 0 & 0 & 0 & 0 & 0 \\
0 & 0 & 1 & 0 & 0 & 0 & 0 & 0 & 0 \\
0 & 0 & 1 & 0 & 0 & 0 & 0 & 0 & 0 \\
0 & 0 & 1 & 0 & 0 & 0 & 0 & 0 & 0
\end{bmatrix}
\begin{bmatrix} 16 \\ \tilde{6} \\ 6 \\ 1 \\ \tilde{1} \\ 1 \\ 1 \\ 1 \\ 1 \end{bmatrix}
\rightarrow
\begin{bmatrix}
0 & 1 & 0 & 1 & 0 & 0 & 0 & 0 & 0 \\
1 & 0 & 1 & 0 & \tilde{1} & 1 & 0 & 0 & 0 \\
0 & 1 & 0 & 0 & 0 & 0 & 1 & 1 & 1 \\
1 & 0 & 0 & 0 & 0 & 0 & 0 & 0 & 0 \\
0 & \tilde{1} & 0 & 0 & 0 & 0 & 0 & 0 & 0 \\
0 & 1 & 0 & 0 & 0 & 0 & 0 & 0 & 0 \\
0 & 0 & 1 & 0 & 0 & 0 & 0 & 0 & 0 \\
0 & 0 & 1 & 0 & 0 & 0 & 0 & 0 & 0 \\
0 & 0 & 1 & 0 & 0 & 0 & 0 & 0 & 0
\end{bmatrix}
$$

$$\text{C-C} \qquad \rightarrow \qquad \text{C-H}$$

Portions of message tapes assembled for the molecules read as

Ethane:

> ...(C-H)(C-H)(C-H)(C-C)(C-H)(C-C)(C-H)(C-H)(C-C)(C-H)(C-H)(C-C)
> (C-H)(C-C)(C-H)(C-C)(C-H)(C-C)(C-H)(C-C)(C-H)(C-H)(C-C)(C-H)
> (C-H)(C-H)(C-C)(C-H)(C-H)(C-H)(C-H)(C-H)(C-H)(C-H)(C-H)
> (C-H)(C-C)(C-H)(C-H)(C-H)(C-C)(C-H)(C-C)(C-H)(C-H)(C-C)(C-H)
> (C-H)(C-C)(C-H)(C-H)(C-C)(C-H)(C-H)(C-H)(C-H)(C-H)(C-C)(C-H)
> (C-C(C-H)(C-H)(C-C)(C-H)(C-H)(C-H)(C-C)(C-H)(C-H)(C-C)(C-H)
> (C-C)(C-H)(C-C)(C-H)(C-H)(C-C)(C-H)(C-H)(C-C)(C-H)(C-H)(C-H)
> (C-H)(C-H)(C-C)(C-H)(C-H)(C-H)(C-H)(C-C)(C-H)(C-H)(C-C)(C-H)...

Propane:

> ...(C-H)(C-H)(C-C)(C-H)(C-H)(C-C)(C-H)(C-C)(C-H)(C-C)(C-H)(C-C)
> (C-H)(C-H)(C-C)(C-H)(C-C)(C-C)(C-H)(C-C)(C-H)(C-H)(C-C)(C-H)
> (C-H)(C-C)(C-H)(C-C)(C-C)(C-H)(C-H)(C-H)(C-H)(C-H)(C-C)(C-H)
> (C-C)(C-H)(C-H)(C-H)(C-H)(C-C)(C-H)(C-C)(C-H)(C-C)(C-H)(C-C)
> (C-H)(C-H)(C-H)(C-H)(C-C)(C-H)(C-C)(C-C)(C-H)(C-C)(C-H)(C-H)
> (C-C)(C-H)(C-C)(C-H)(C-C)(C-H)(C-C)(C-H)(C-C)(C-H)(C-H)(C-C)
> (C-H)(C-H)(C-C)(C-H)(C-H)(C-H)(C-C)(C-H)(C-H)(C-H)(C-H)(C-C)
> (C-H)(C-H)(C-C)(C-H)(C-C)(C-H)(C-C)(C-C)(C-H)(C-H)(C-C)(C-H)...

Ethanethiol:

> ...(C-H)(C-C)(C-S)(C-H)(C-S)(S-H)(C-S)(C-C)(C-H)(C-C)(C-H)(C-H)
> (C-C)(C-S)(S-H)(C-S)(C-H)(C-H)(C-C)(C-H)(C-C)(C-H)(C-H)(C-C)
> (C-H)(C-H)(C-H)(C-H)(C-H)(C-H)(C-C)(C-H)(C-H)(C-S)(S-H)(C-S)
> (C-H)(C-S)(C-C)(C-H)(C-H)(C-H)(C-C)(C-H)(C-C)(C-H)(C-H)(C-H)
> (C-H)(C-C)(C-H)(C-H)(C-H)(C-C)(C-H)(C-H)(C-H)(C-H)(C-C)(C-H)
> (C-S)(S-H)(C-S)(C-H)(C-S)(C-H)(C-H)(C-H)(C-S)(S-H)(C-S)(C-H)
> (C-C)(C-H)(C-H)(C-C)(C-S)(S-H)(C-S)(C-H)(C-S)(S-H)(C-S)(C-H)
> (C-C)(C-H)(C-H)(C-H)(C-H)(C-C)(C-H)(C-S)(S-H)(C-S)(S-H)(C-S)...

Each tape is unique due to the electronic composition and structure of the source. Each tape is without limit, at least in principle, because there is no end to the collisions in a heat-filled environment. Practically speaking, it is straightforward to compile tapes of several thousand ABA units or more. Appendix A of this book presents a computer program that will perform such compilations. This program can be adapted to a variety of small organic molecules to probe their information properties. The lesson of Figure 6.2 must be kept in mind, however, that every site of a molecule offers multiple interactions. Thus, each entry C-H, C-C, and so forth of a record tape stands for a set of electronic messages. It is the sequence of units that contains information in a manner that parallels polypeptides and polynucleotides: ...RVRRVRRV... and ...CUCGACGU.... As with biopolymers, the information for small compounds can be quantified in bits at multiple orders: first, second, third, and so forth. Each order corresponds to the number of message units transmitted and registered in a Brownian process.

Ethane and propane are restricted to C-H and C-C units. The same units are carried by ethanethiol, in addition to C-S and S-H. To quantify information in the first order—the bits per single ABA encounter—one tabulates the occurrence frequencies (f_i) for each unit recorded on the tape. A given molecule poses N different units. The first-order Shannon information (I_1) follows from the (now) familiar formula:

$$I_1 = -\sum_i^N f_i \cdot \log_2 f_i \tag{6.11}$$

Therefore for ethane and propane:

$$I_1 = -f_{C-H} \cdot \log_2 f_{C-H} - f_{C-C} \cdot \log_2 f_{C-C} \tag{6.12}$$

whereas for ethanethiol, one computes:

$$I_1 = -f_{C-H} \cdot \log_2 f_{C-H} - f_{C-C} \cdot \log_2 f_{C-C} - f_{C-S} \cdot \log_2 f_{C-S} - f_{S-H} \cdot \log_2 f_{S-H} \tag{6.13}$$

The first-order information for acetone would follow from the three-term expression:

$$I_1 = -f_{C-H} \cdot \log_2 f_{C-H} - f_{C-C} \cdot \log_2 f_{C-C} - f_{C=O} \cdot \log_2 f_{C=O} \tag{6.14}$$

It is straightforward to identify via the formula diagram the terms necessary for computing I_1, as is the subject of a few end-of-chapter exercises. It is not always so easy to anticipate f_i in advance of the message tape. For example, C-C constitutes one of seven ABAs in ethane. However, its tape frequency is 0.250 on account of extended collisions and nearest-neighbor effects. If a thermal contact transpires at C-H, there is a one in four chance that the next (i.e., influential) message unit will be C-C.

For two-unit sequences, ethane manifests three possibilities: (C-H)(C-C), (C-C)(C-H), and (C-H)(C-H). Thus Shannon information arrives in the second order via:

$$I_2^{ethane} = -f_{(C-H)(C-C)} \cdot \log_2 f_{(C-H)(C-C)} - f_{(C-C)(C-H)} \cdot \log_2 f_{(C-C)(C-H)} - f_{(C-H)(C-H)} \cdot \log_2 f_{(C-H)(C-H)}$$

$$\tag{6.15}$$

Molecules such as propane, butane, pentane, and so forth offer the same pair combinations as ethane plus (C-C)(C-C). Ethanethiol poses the same pair sequences as ethane along with (C-H)(C-S), (C-S)(C-H), (C-S)(S-H), (S-H)(C-S), (C-C)(C-S), and (C-S)(C-C). Accordingly there are nine terms to address for $I_2^{ethanethiol}$. Counting the pair and triplet messages for acetic acid and acetone is the subject of an exercise. Clearly the number of messages grows with the complexity of the molecule and the order (length) of the contact sequence. It is straightforward to identify examples of high-order messages by looking at the formula diagram and imagining the random walk. For instance, ethanethiol offers the following in the sixth order:

...(S-H)(C-S)(C-H)(C-H)(C-H)(C-S)...
...(C-H)(C-C)(C-H)(C-H)(C-H)(C-H)...
...(C-H)(C-H)(C-H)(C-H)(C-C)(C-S)...

It is by no means easy to write all the possibilities by visual inspection. Rather the possible thermal walks must be fleshed out with the help of computer programs and then parsed for all multiorder messages.

Figure 6.5 illustrates the Shannon information allied with ethane, propane, and ethanethiol as a function of sequence order n. Vertical error bars mark the averages ±1 standard deviation; the bars are barely perceptible given the symbol sizes. The errors arise from the finiteness of the message tapes and drifts in the random number generator.

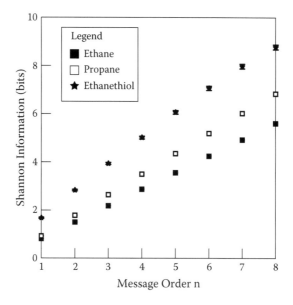

FIGURE 6.5   Shannon information for ethane, propane, and ethanethiol. Error bars mark averages ±1 standard deviation. Information values derive from the statistical structure of the message sequences allowed by the formula diagrams. The abscissa marks the number of message units.

The results make several points. First, the more diverse the ABA composition, the greater the Shannon information. At all orders for the three molecules:

$$I_n^{ethanethiol} > I_n^{propane} > I_n^{ethane} \qquad (6.16)$$

The data also convey that the information availed by a molecule increases with the number of message units transmitted and registered. For any compound:

$$I_{n-1} < I_n < I_{n+1} \qquad (6.17)$$

Figure 6.5 also shows the Shannon information to be linear with the sequence order. This leads to a state descriptor for the microscopic scale based on the natural rise of I_n with n. In particular, linear regression analyses lead to a best-fit slope ξ_I with uncertainty σ_ξ, each having units of bits per message unit. For the three Bunsen burner molecules, one finds:

$$\xi_I^{ethane} = 0.686, \quad \sigma_{\xi,I}^{ethane} = 5.9 \times 10^{-4} \text{ bits/message unit}$$

$$\xi_I^{propane} = 0.846, \quad \sigma_{\xi,I}^{propane} = 3.5 \times 10^{-3} \text{ bits/message unit}$$

$$\xi_I^{ethanethiol} = 1.02, \quad \sigma_{\xi,I}^{ethanethiol} = 2.2 \times 10^{-2} \text{ bits/message unit}$$

All systems pose more than one type of information. For molecules, the code units manifest frequencies such as f_{C-H} and f_{C-C} in thermal collision strings. A record tape that logs the collisions also offers pair combinations, for example, C-H followed immediately by C-C at frequency $f_{(C-H)(C-C)}$. It is the mutual information (MI) that quantifies the correlations imbedded in these unit sequences. For instance, ethane offers three message pairs. MI arrives in the second order of analysis via:

$$MI_2^{ethane} = +prob\big((C-C)(C-H)\big) \cdot \log_2\left(\frac{prob\big((C-C)(C-H)\big)}{prob(C-C) \cdot prob(C-H)}\right)$$

$$+ prob\big((C-H)(C-C)\big) \cdot \log_2\left(\frac{prob\big((C-H)(C-C)\big)}{prob(C-H) \cdot prob(C-C)}\right) \qquad (6.18)$$

$$+ prob\big((C-H)(C-H)\big) \cdot \log_2\left(\frac{prob\big((C-H)(C-H)\big)}{prob(C-H) \cdot prob(C-H)}\right)$$

The computations for MI_2 for propane and ethanethiol feature four and nine terms, respectively. The correlations within a message tape are not limited to pairs. Thus, the mutual information expressed by a tape can be quantified at multiple correlation orders 3, 4, 5, …. For instance, one of the third-order terms for ethane is:

$$MI_{3,(C-C)(C-H)(C-C)}^{ethane} = +prob((C-C)(C-H)(C-C)) \cdot \log_2 \left(\frac{prob((C-C)(C-H)(C-C))}{prob(C-C) \cdot prob(C-H) \cdot prob(C-C)} \right)$$

(6.19)

and an example of a fourth-order term is:

$$MI_{4,(C-C)(C-H)(C-H)(C-C)}^{ethane} = +prob((C-C)(C-H)(C-H)(C-C))$$

$$\times \log_2 \left(\frac{prob((C-C)(C-H)(C-H)(C-C))}{prob(C-C) \cdot prob(C-H) \cdot prob(C-H) \cdot prob(C-C)} \right)$$

(6.20)

Such would carry zero weight if the message units were logged independent of one another on a record tape. *MI* quantities are illustrated in Figure 6.6 for ethane, propane, and ethanethiol. As with Shannon information, *MI* increases with correlation order *n* in a linear fashion. Striking is the contrast between the hydrocarbons and the organosulfur compound. One thus obtains a second descriptor of the Angstrom-scale

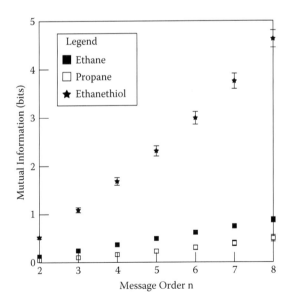

FIGURE 6.6 Mutual information for ethane, propane, and ethanethiol. Error bars mark averages ±1 standard deviation. Quantities are based on the statistical structure of the formula diagrams subject to a nearest-neighbor random walk.

state. Linear regression analyses of MI_n versus n lead to:

$$\xi_{MI}^{ethane} = 0.125, \quad \sigma_{\xi,MI}^{ethane} = 6.5 \times 10^{-4} \text{ bits/message unit}$$

$$\xi_{MI}^{propane} = 0.0741, \quad \sigma_{\xi,MI}^{propane} = 3.6 \times 10^{-3} \text{ bits/message unit}$$

$$\xi_{MI}^{ethanethiol} = 0.677 \quad \sigma_{\xi,MI}^{ethanethiol} = 2.3 \times 10^{-2} \text{ bits/message unit}$$

Earlier chapters showed that a system can be described macroscopically by energy quantities such as U and H. The record tape for a molecule makes contact with the Angstrom scale by reference to ABA energy tables. For example, ethanethiol hosts four message units: C-H, C-C, C-S, and S-H. The average energy $< D_1 >$ in the first order obtained from computing is:

$$\langle D_1 \rangle = f_{C-H} \cdot D_{C-H} + f_{C-C} \cdot D_{C-C} + f_{C-S} \cdot D_{C-S} + f_{S-H} \cdot D_{S-H} \tag{6.21}$$

The average energy in the second order follows from the expression:

$$\langle D_2 \rangle = f_{(C-H)(C-C)} \cdot (D_{C-H} + D_{C-C}) + f_{(C-H)(C-H)} \cdot (D_{C-H} + D_{C-H})$$
$$+ f_{(C-C)(C-S)} \cdot (D_{C-C} + D_{C-S}) + \cdots \tag{6.22}$$

There are six additional terms to calculate and the expressions for $< D_3 >$, $< D_4 >$, and so forth follow in like fashion.

Energy computations lead to Figure 6.7 based on the message tapes for ethane, propane, and ethanethiol. Not surprisingly, $<D_n>$ increases linearly with order n, although there is little contrast among the molecules. A third descriptor arrives:

$$\xi_{<D>}^{ethane} = 397, \quad \sigma_{\xi,<D>}^{ethane} = 5.5 \times 10^{-4} \text{ kilojoules/mole} \cdot \text{message unit}$$

$$\xi_{<D>}^{propane} = 391, \quad \sigma_{\xi,<D>}^{propane} = 3.5 \times 10^{-3} \text{ kilojoules/mole} \cdot \text{message unit}$$

$$\xi_{<D>}^{ethanethiol} = 366, \quad \sigma_{\xi,<D>}^{ethanethiol} = 1.8 \times 10^{-3} \text{ kilojoules/mole} \cdot \text{message unit}$$

These follow from linear regression analyses applied to $<D_n>$ versus n.

At the macroscopic level, a system offers energy dispersion quantities such as σ_U and σ_H. Molecular message tapes do their part by furnishing σ_D at multiple orders. In a first-order analysis of ethanethiol tapes, one computes:

$$\sigma_D^{(1)} = \sqrt{(f_{C-H} \cdot D_{C-H} - \langle D_1 \rangle)^2 + (f_{C-C} \cdot D_{C-C} - \langle D_1 \rangle)^2 + (f_{C-S} \cdot D_{C-S} - \langle D_1 \rangle)^2 + (f_{S-H} \cdot D_{S-H} - \langle D_1 \rangle)^2}$$

$$\tag{6.23}$$

In the second order, one calculates:

$$\sigma_D^{(2)} = \sqrt{(f_{(C-H)(C-C)} \cdot (D_{C-H} + D_{C-C}) - \langle D_2 \rangle)^2 + (f_{(C-H)(C-H)} \cdot (D_{C-H} + D_{C-H}) - \langle D_2 \rangle)^2 + \cdots}$$

$$\tag{6.24}$$

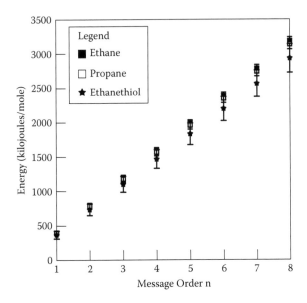

FIGURE 6.7   ABA energies allied with message tapes. Error bars mark averages ±1 standard deviation. Quantities are based on the message tapes compiled via random walks and reference to standard tables of dissociation energies.

Dispersion quantities have been plotted in Figure 6.8. The scaling is not as linear as in the three preceding figures. At the same time, there is greater contrast among the molecules. A fourth descriptor of the microscopic state follows from linear regression analyses:

$$\xi^{ethane}_{\sigma_D} = 4.57, \quad \sigma^{ethane}_{\xi,\sigma_D} = 0.18 \text{ kilojoules/mole} \cdot \text{message unit}$$

$$\xi^{propane}_{\sigma_D} = 5.81, \quad \sigma^{propane}_{\xi,\sigma_D} = 0.28 \text{ kilojoules/mole} \cdot \text{message unit}$$

$$\xi^{ethanethiol}_{\sigma_D} = 21.3, \quad \sigma^{ethanethiol}_{\xi,\sigma_D} = 1.0 \text{ kilojoules/mole} \cdot \text{message unit}$$

A sample of ethane, propane, or ethanethiol can be described macroscopically using n, V, T, p, and other state variables. It is the ξ values that furnish elementary descriptors of states at the microscopic level.

6.3   CHARACTERISTICS OF INFORMATION AT THE MOLECULAR LEVEL

Chapter 3 demonstrated that thermodynamic systems afford substantive information in the statistical sense when they are very small in size. This is echoed by molecules. Substantive information manifests at the Angstrom level because a random process—thermal motion plus collisions—is always in place for message transmission and registration. Chapter 4 discussed how information is realized when a structured

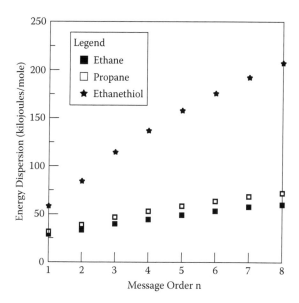

FIGURE 6.8 ABA dispersion energies allied with message tapes. Quantities are based on the message tapes and standard tables of dissociation energies.

program is applied to a system—a program demarcates a state population that is both extended and diverse. Molecules have programs built into them, so to speak, by their element composition and covalent bond network. Thermal energy does the rest in powering translations, rotations, and collisions.

The second characteristic concerns standard tables. These provide indispensable data for the chemist regarding molar entropies, enthalpies, free energies, and ABA energies. Concerning information, the tables can be augmented by applying the methods of the previous section. Table 6.1 lists $\xi_I, \xi_{MI}, \xi_{<D>}, \xi_{\sigma_D}$ for familiar compounds. Clearly, the greatest variations are demonstrated by the correlation and dispersion quantities ξ_{MI} and ξ_{σ_D}; with rare exception, the least variations are found in $\xi_{<D>}$. It is challenging to anticipate ξ descriptors prior to message tape assembly and analysis. The reason is that the ABA sequences depend intricately on the source structures and random walk properties. This should not be surprising. It is difficult to intuit the statistical structure of information-bearing molecules in general; proteins and polynucleotides offer plenty of examples here.

The third characteristic concerns the bits per message unit encountered—it is typically low for an organic molecule. The largest ξ_I in Table 6.1 are allied with 2-chlorocyclopentanone and 1,2-dibromobutane at 1.16 bits per message unit. The electronic programs of molecules demonstrate an economy of information.

Chapters 4 and 5 illustrated special transformations marked by zero information in one or more state variables. Regarding the Angstrom level, the fourth characteristic is the specialness of certain compounds. Table 6.1 reports $\xi_I, \xi_{MI}, \xi_{\sigma_D}$ for CH_4, N_2, O_2, CO_2, and H_2O as zero. These charge packages are special because they lack ABA diversity; elements packaged in nature as molecules—Cl_2, F_2, Br_2, and so forth—are special in this respect. Nonzero information for these systems is

TABLE 6.1

Angstrom-Scale Descriptors of Various Molecules

Molecule	ξ_I	ξ_{MI}	$\xi_{<D>}$	ξ_{σ_D}
Arginine	1.12	1.53	368	26.6
1-Butene	1.07	0.324	421	27.8
2-Chloro-cyclopentanone	1.16	0.332	389	16.0
Cyclopentanone	1.04	0.185	393	15.4
1,2-Dibromobutane	1.16	0.275	369	16.3
Ethane	0.685	0.125	397	4.57
Ethanethiol	1.022	0.677	366	21.3
n-Butane	0.889	0.0683	388	6.58
n-Propane	0.846	0.0741	391	5.81
iso-Butane	0.886	0.0765	388	7.18
Valine	1.06	1.25	388	20.8
CO_2	0.00	0.00	802	0.00
O_2	0.00	0.00	498	0.00
CH_4	0.00	0.00	414	0.00
H_2O	0.00	0.00	464	0.00

Note: ξ_I and ξ_{MI} are listed in units of bits per atom–bond–atom (ABA) unit. $\xi_{<D>}$ and ξ_{σ_D} are listed in kilojoules per mole per ABA unit.

registered only at lower levels, such as by taking into account the resonance and atomic orbital structures. Electronic diversity in the molecules is not captured in stand-alone digital reductions provided by ABA units.

The fifth characteristic recognizes that the mutual information is not zero at the ABA level, the exceptions being the molecules of the preceding paragraph. This is on account of the electronic structure correlations. If nature assembled molecules absent valence and other structure rules, ξ_{MI} would forever be pinned at zero. Among other consequences, a collision at one site of a compound would afford no information about a neighbor. Note that the ξ_{MI} of Table 6.1 span more than an order of magnitude. It is the mutual information of a molecule's messages that provide a signature attribute.

Additional characteristics are illustrated by transforming and replotting the Table 6.1 data. Figures 6.9 through 6.11 follow from Table 6.1 entries using the data for ethane as a baseline. Illustrated are reduced (dimensionless) descriptors, applying the identical scale to the horizontal and vertical axes. For example, the reduced descriptor for the Shannon information is obtained from:

$$\hat{\xi}_I^{1-butene} = \frac{\xi_I^{1-butene}}{\xi_I^{ethane}} \qquad (6.25)$$

Reduced forms of $\xi_{MI}, \xi_{<D>}$ and ξ_{σ_D} follow in the analogous way. The point allied with each molecule has been labeled. The error bars are established by the $\sigma_{\xi,I}$ and $\sigma_{\xi,<D>}$ from regression analyses.

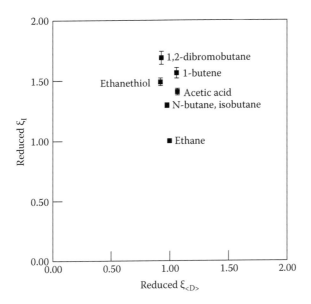

FIGURE 6.9 Reduced descriptors for two- and four-carbon systems. Plotted are $\hat{\xi}_I$ versus $\hat{\xi}_D$ scaled via ethane data. The point for ethane molecule appears at 1,1. The error bars are established by $\sigma_{\xi,I}$ and $\sigma_{\xi,<D>}$ from regression analyses.

Figures 6.9 and 6.10 concentrate on two- and four-carbon systems and carry important lessons. The purpose of a functional group is to activate a molecule in a spatially selective way; the group lowers the chemical inertness at one or a few sites. These second-year chemistry ideas are reinforced by the information characteristics. Figure 6.9 shows how the Shannon information is enhanced by 50% by functionalizing ethane using either a carboxyl or mercapto group. Note the vertical alignment of the data. This tells us that the more prominent effect of functionalizing a molecule entails information, not energy density. Likewise approximately 40% of the enhancements are demonstrated by converting *n*-butane into an alkene or dibromo-derivative; the energy changes per ABA unit are incidental by contrast. A functional group avails new electronic programs in a molecule for controlling the work and heat transactions of chemical reactions. It does so by enhancing the carrier's information of the statistical variety. Figure 6.10 makes an analogous point regarding energy dispersion. Here the effects of functional groups are even more pronounced, for example, a mercapto group enhances $\hat{\xi}_{\sigma_D}$ by nearly a factor of five. Note as well that $\hat{\xi}_{\sigma_D}$ slightly discriminates *normal-* from *iso-*butane owing to the differences in their electronic structure. The added point is that functional groups amplify the energy diversity of the electronic message space.

Figure 6.11 illustrates state points in the $\hat{\xi}_{MI}, \hat{\xi}_{<D>}$ plane for all but the special molecules of Table 6.1. A sharp distinction is demonstrated again between activated molecules and those lacking functional groups. It is the mutual information that is indeed most sensitive to chemical modification. Activating ethane with a carboxyl group increases the collision-based Shannon information by 50%; the enhancement

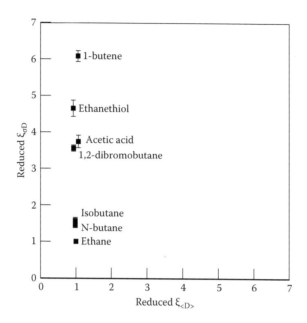

FIGURE 6.10 More state descriptors for two- and four-carbon systems. Plotted are $\hat{\xi}_{\sigma_D}$ versus $\hat{\xi}_D$ scaled via ethane data. The point for ethane appears at 1,1 in the state plane.

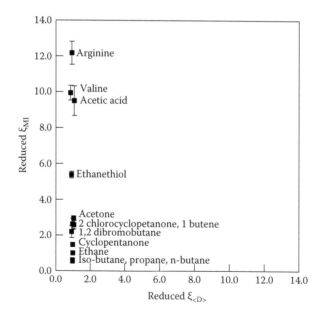

FIGURE 6.11 Reduced descriptors for Table 6.1 molecules. Plotted are $\hat{\xi}_{MI}$ versus $\hat{\xi}_{\sigma_D}$ scaled via ethane data. The point for ethane appears at 1,1.

is by nearly a factor of 10 concerning mutual information. The figure makes the additional point that amino acids (cf. Figure 2.2 of Chapter 2) are especially endowed with message space correlations. As is well appreciated, these compounds are prominent in biochemical signaling and reaction control, as opposed to saturated hydrocarbons. One looks to peptides and proteins for showcase examples.

The major points of this chapter are the following:

1. Systems express information not just at the macroscopic level. There are robust states, mechanisms for communication and registration, and uncertainty at the microscopic scale as well. This scale is the complicated domain of quantum mechanics and statistical thermodynamics. Standard tables, formula diagrams, and random walk models, however, provide more immediate handles on information in the statistical sense.

2. The statistical structure of a molecule and message sequences render state descriptors $\xi_I, \xi_{MI}, \xi_{<D>}$ and ξ_{σ_p} as in Table 6.1. Of these, ξ_{MI} and ξ_{σ_D} are the most differentiated and indeed fingerprint-like for molecules. Amino acids, alkaloids, and natural products demonstrate highest ξ_{MI}, whereas saturated hydrocarbons express the lowest. One is reminded how information plays more than one role for a system and its environment. It equates with the code amounts needed for labeling states or messages; it connects with a system's diversity, complexity, and capacity for control. It is then no surprise that functionalized molecules such as acids and bases express greater information on Shannon and mutual accounts compared with alkanes. Molecules especially endowed with functional groups (e.g., amino acids) are able to control a greater number of chemical decisions, all ultimately having to do with the transfer of work and heat.

6.4   SOURCES AND FURTHER READING

The intersection of the microscopic scale with information presents a vast literature. To list a sampling most helpful to the author, one begins with the information theory and statistical thermodynamics work of Jaynes [4], and the later text by Baierlein on atoms and information [5]. At a less advanced but still highly illuminating level are books by Morowitz [6,7]. Information casts a wide net in chemistry. Levine and coworkers have long championed information theory applied to molecular processes such as relaxation and internal energy redistribution [8,9]. Biopolymers plus information yield the field of bioinformatics. Recommended is the text by Tramontano for the landmark questions posed [10]. The research of Schneider has addressed in depth the information attributes of biopolymers [11,12].

Regarding microscopic state descriptors, the stage was set in the late 1970s by the work of Bonchev and Trinajstic on the branching processes of alkanes [13]. Included in this reference are key sources regarding the information posed by vertex graphs, the standard vehicles for portraying organic compounds. Along related lines, the complexity of molecules via their structural information was formalized in the 1980s by Bertz [14,15]. Several research groups have approached molecular information by quantifying topological indices of an entropic nature [16–18]. González-Díaz and coworkers have pioneered the application of Markov and entropic descriptors of numerous molecular systems [19]. The molecular message tapes illustrated in this chapter share many properties with Markov chains.

The information can be studied for individual molecules or libraries. Bajorath and coworkers have established signature contrasts between natural and synthetic libraries [20,21]. For several years, the author and his students have investigated molecular information at the base formula and structure diagram levels. The Brownian computation model described in Chapter 6 has been directed to individual molecules, libraries, and enzymatic proteins [22–25].

There are three more notes to add. Bennett discussed more than just molecular information processing; he considered as well the principles of mechanical computation [3]. The fundamentals have received further elaboration in the Feynman lectures on computation [26]. Second, as indicated in Chapter 1, ABA units encode molecular information at a high level. Information at a deeper level has been explored extensively by the work of Parr and Yang [27] and Nalewajski [28]. Last, thermochemical tables are indispensable to all branches of chemistry. Extensive compilations have been presented by Cox and Pilcher [29].

6.5   SUGGESTED EXERCISES

The following require adapting the program listed in Appendix A to diverse molecules. A variety of compounds can be compared and contrasted via information descriptors.

6.1   Table 6.1 lists ξ_{MI} for acetic acid as significantly greater than ξ_{MI} for ethane. Does esterification of the acid via ethyl alcohol enhance or diminish ξ_{MI}? Please discuss in terms of information as a type of control capacity.

6.2   How do ξ_{MI} and ξ_{σ_D} compare for cyclobutane (left) and cubane (right)? Please discuss. Were the results anticipated correctly?

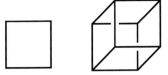

6.3   How do ξ_{MI} and ξ_{σ_D} compare for cyclobutene (left) and Dewar benzene (right)? Please discuss.

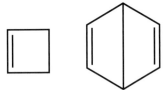

6.4   The *Handbook of Chemistry and Physics* lists five compounds with formula C_6H_{14}. Review the structures and posit which offers the largest ξ_{MI}. Do likewise regarding ξ_{σ_D}. Check the prediction by constructing and analyzing message tapes for the molecules. Were the results anticipated correctly in advance?

6.5 How do ξ_{MI} and ξ_{σ_D} compare for cyclohexane (left) and adamantane (right)? Please discuss.

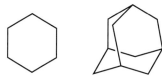

6.6 How do ξ_{MI} and ξ_{σ_D} compare for leucine and isoleucine? Please discuss.

6.7 A famous experiment involved mixtures of methane, ammonia, water, and hydrogen subject to electrical sparks. The reaction products included the amino acids glycine, aspartic acid, glutamic acid, and β-alanine in mole ratios 63:0.4:0.6:15. How do the values of $\xi_I, \xi_{MI}, \xi_{<D>}$ and ξ_{σ_D} correlate with these results? The chemistry is discussed at length by Calvin [30].

6.8 In early days, camphor (right) was obtained via reactions commencing with α-pinene (left). In carrying out the synthesis, does the chemist enhance or diminish the mutual information expressed by the starting material?

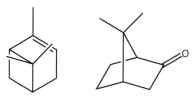

6.9 Consider the thermal collision messages of molecules without the help of computer programs. Identify all the first-, second-, and third-order messages for acetone. Do likewise for acetic acid.

REFERENCES

[1] The molecules in natural gas have been of interest to fuel researchers for more than a century. For ground-floor studies, see Burrell, G. A., Siebert, F. M. 1915. The Separation of the Constituents in a Natural Gas from Which Gasoline Is Condensed, *J. Amer. Chem. Soc.* 37, 392.
[2] Lehninger, A. L. 1970. *Biochemistry*, chap. 31, Worth Publishers, New York.
[3] Bennett, C. H. 1982. Thermodynamics of Computation—A Review, *Intl. J. Theo. Phys.* 21, 905.
[4] Jaynes, E. T. 1957. Information Theory and Statistical Mechanics, *Phys. Rev.* 106, 620; Information Theory and Statistical Mechanics II, *Phys. Rev.* 108, 171.
[5] Baierlein, R. 1971. *Atoms and Information Theory: An Introduction to Statistical Mechanics*, Freeman, San Francisco.
[6] Morowitz, H. J. 1970. *Energy for Biologists: An Introduction to Thermodynamics*, Academic Press, New York.

[7] Morowitz, H. J. 1979. *Energy Flow in Biology: Biological Organization as a Problem in Thermal Physics*, Ox Bow Press, Woodbridge, CT.

[8] Levine, R. D. 1978. The Information Theoretic Approach to Molecular Reaction Dynamics, *Annu. Rev. Phys. Chem.* 29, 59.

[9] Levine, R. D. 1981. The Information Theoretic Approach to Intramolecular Dynamics, *Adv. Chem. Phys.* 47, 239.

[10] Tramontano, A. 2005. *The Ten Most Wanted Solutions in Protein Bioinformatics*, Chapman & Hall/CRC, Boca Raton, FL.

[11] Schneider, T. D. 1991. Theory of Molecular Machines. I. Channel Capacity of Molecular Machines, *J. Theor. Biol.* 148, 83; II. Energy Dissipation from Molecular Machines, *J. Theor. Biol.* 148, 125.

[12] Schneider, T. D. 1997. Information Content of Individual Genetic Sequences, *J. Theor. Biol.* 189, 427.

[13] Bonchev, D., Trinajstic, N. 1978. Information Theory, Distance Matrix, and Molecular Branching, *J. Chem. Phys.* 67, 4517.

[14] Bertz, S. H. 1981. The First General Index of Molecular Complexity, *J. Amer. Chem. Soc.* 103, 3599.

[15] Bertz, S. H., Herndon, W. C. 1986. The Similarity of Graphs and Molecules in American Chemical Society, Washington, D.C. *Artificial Intelligence Applications in Chemistry*, chap. 15, 169–175. DOI: 10.1021/bk-1986–0306. ch015.

[16] Basak, S. C., Mekenyan, O. 1994. Topological Indices and Chemical Reactivity, in *Graph Theoretical Approaches to Chemical Reactivity*, Bonchev, D., Mekenyan, O., eds. chap. 8., Kluwer Academic, Dordrecht.

[17] Basak, S. C., Grunwald, G. D., Niemi, G. J. 1997. Use of Graph-Theoretic and Geometrical Molecular Descriptors in Structure-Activity Relationships, in *From Chemical Topology to Three-Dimensional Geometry*, Balaban, A. T., ed., chap. 4, Plenum Press, New York.

[18] Bonchev, D. 1993. *Information Theoretic Indices for Characterization of Chemical Structure*, Research Studies Press–Wiley, Chichester.

[19] de Ramos, A. R., González-Díaz, H., Molina, R. R., Uriarte, E. 2004. Markovian Backbone Negentropies: Molecular Descriptors for Protein Research. I. Predicting Protein Stability in Arc Repressor Mutants, *Proteins: Struct. Func. Bioinf.* 56, 715.

[20] Godden, J. W., Stahura, F. L., Bajorath, J. 2000. Variability of Molecular Descriptors in Compound Databases Revealed by Shannon Entropy Calculations, *J. Chem. Inf. Comput. Sci.* 40, 796.

[21] Stahura, F. L., Godden, J. W., Xue, L., Bajorath, J. 2000. Distinguishing Between Natural Products and Synthetic Molecules by Descriptor Shannon Entropy Analysis and Binary QSAR Calculations, *J. Chem. Inf. Comput. Sci.* 40, 1245.

[22] Graham, D. J., Malarkey, C., Schulmerich, M. V. 2004. Information Content in Organic Molecules: Quantification and Statistical Structure via Brownian Processing, *J. Chem. Inf. Comput. Sci.* 44, 1601.

[23] Graham, D. J. 2005. Information Content and Organic Molecules: Aggregation States and Solvent Effects, *J. Chem. Inf. Modeling* 45, 1223.

[24] Graham, D. J. 2007. Information Content in Organic Molecules: Brownian Processing at Low Levels, *J. Chem. Inf. Modeling* 47, 376.

[25] Graham, D. J., Greminger, J. L. 2009. On the Information Expressed in Enzyme Primary Structure: Lessons from Ribonuclease A, *Molecular Diversity*. DOI: 10.1007/s11030-009-9211-3.

[26] Feynman, R. P. 1996. *Feynman Lectures on Computation*, A. J. G. Hey, R. W. Allen, eds., Addison-Wesley, Reading, MA.

[27] Parr, R. G., Yang, W. 1989. *Density-Functional Theory of Atoms and Molecules*, Oxford University Press, Oxford.

[28] Nalewajski, R. F. 2006. *Information Theory of Molecular Systems*, Elsevier, Amsterdam.

[29] Cox, J. D., Pilcher, G. 1970. *Thermochemistry of Organic and Organometallic Compounds*, Academic Press, New York.

[30] Calvin, M. 1969. *Chemical Evolution*, chap. 6, Oxford University Press, New York.

7 Thermodynamic Information and Chemical Reactions

The states of a system are modified by variable tuning and energy exchanges. In chemically active venues, the states alter spontaneously and with purpose in collaboration with the surroundings. This chapter considers an unusual type of thermodynamic transformation by way of chemical reactions.

7.1 OVERVIEW OF CHEMICAL REACTIONS

In Chapter 3, the states of a system were specified by p, V, and other variables. It was shown that information in the statistical sense was low in most cases and indeed bordered on zero. The reason is that fluctuations wield only tiny impacts for large volume, multiparticle systems under equilibrium conditions. Matters are different when structured programs are applied. All the programmed pathways of Chapters 4 and 5 featured extended collections of states. For a given collection, there was appreciable information allied with the variables in query-and-measurement exercises. The exceptions were n for closed systems, and p, T, S, and so forth for isobaric, isothermal, and adiabatic—the special transformations of thermodynamics.

Chapter 6 turned to the microscopic level. All molecules underpin probability functions via their charge distributions. Thermal environments do their part by imposing uncertainty on all electronic communication and registration. It was shown that familiar compounds—ethane, propane, and ethanethiol, for example—pose collision-based information. The atom–bond–atom (ABA) units of the molecules furnish a robust code for labeling the messages. The Shannon and mutual information were quantified for the collision sequences allowed by the molecular structure. This Brownian approach offered new descriptors of states by way of ξ_I, ξ_{MI}, $\xi_{<D>}$, and ξ_{σ_D}.

As the chemical enterprise demonstrates, molecules are valuable not only for what they are, but also for what they can become. Molecule A can beget B and vice versa in environments ranging from flames to cells to round-bottom flasks. The transformations are described in the most succinct terms:

$$A \rightleftharpoons B$$

Such transformations conserve mass, charge, energy, and atom identity. At the same time, they represent a most unusual consequence of thermodynamic fluctuations

and program applications. A discussion of this topic typically concentrates on standard state functions $\Delta G_f^o, \Delta H_f^o$, and S^o; these were briefly visited for Bunsen burner reactants and products of the previous chapter. We shall aim here, however, for a more general and intuitive grasp. This arrives by examining a few cases of reaction thermodynamics from complementary points of view. The major points echo ideas discussed in Chapter 3 for composite systems.

To begin, a thumbnail sketch of why chemical reactions happen goes as follows. If A offers any mechanism for entropy increases, then such increases will occur sooner or later subject to the constraints in place. Left in the wake will be compound B with a definitive structure of its own, and a solution mixture of A and B. The initiation involves contact of A with its neighbors—other A molecules or activating sites—and energy exchanges with the surroundings.

The longer story, of course, involves electronic messages, thermal collisions, and adjustments of the system composition. This is the subject of Figure 7.1, which represents in schematic terms a gas coupled to heat and work reservoirs. For discussion purposes, let the container be leak-proof and held at constant temperature and pressure. Let the system be composed only of A initially as represented by the open circles. Let one of the A molecules be poised for conversion to B, the latter represented by the filled circle.

Now the ordinary events inside a gas sample surround collisions and energy redistribution. The record of any single event is quickly lost as the colliding parties separate and move toward neighbors. The system entropy is diminished slightly during each contact, yet is restored via subsequent collisions. The average entropy remains at some baseline value if no irreversibility occurs anywhere in the event chain.

But reactions are extraordinary events. When A converts to B and repositions the system closer to equilibrium, there occurs a material entry in the record of messages

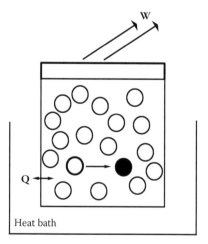

FIGURE 7.1 A chemically reactive gas coupled to heat and work reservoirs. A molecules are represented by the open circles. A single A is poised for conversion to B, the latter represented by the filled circle.

transmitted and registered. The effects can only be nullified at a cost of work imported from outside. The consequences include an increase in the total entropy; this is dispersed unevenly across the system and surroundings. Any restoration to the initial value also carries a work price tag. It is often possible for the chemist to force B back to A and thus lower the entropy. Yet the requisite work has to originate from a natural (i.e., spontaneous) process, which itself can only be reversed via work drawn from yet another source—on and on. The point is that when chemical reactions proceed under nonequilibrium conditions, there is a definite irreversibility in their character. The effects propagate well beyond the site where A converts to B.

Charge distributions underpin messages and communication mechanisms. They also dictate the energy mismatches among molecules. It is typical that the energy—both kinetic and potential—of A's electrons and nuclei does not equate with B's. By the first law of thermodynamics, energy must be conserved in any process, reversible and otherwise. In conversions of A to B, energy shortfalls must be covered by whatever surrounds A and B. The first law requires:

$$E_{transferred\ from\ surroundings} + E_A = E_B \tag{7.1}$$

By the same token, energy windfalls must be absorbed. The first law also mandates:

$$E_A = E_B + E_{transferred\ to\ surroundings} \tag{7.2}$$

The most common currency is heat propagated by collisions. In the system of Figure 7.1, any transfer of heat forces an entropy change in the surroundings, namely,

$$\Delta S_{surroundings} = \frac{-Q_{rec}}{T} \tag{7.3}$$

The negative sign is critical. For an endothermic process, heat is transferred from the bath (heat reservoir) to A. The entropy of the reservoir is thereby diminished. For an exothermic process, heat must diffuse outward toward the bath whereby its entropy increases. Since the discussion restricts the heat exchanges to constant T, p conditions, they (i.e., Q_{rec} values) connect simply with changes in the system enthalpy.

The entropy effects are not confined to the bath. Even if the production of B is highly local, such as at an activating or catalytic site, there will be eventual and thorough mixing of the reactants and products. The chemist would have to expend work should he or she need to separate A from B. Let the mixing entropy for n_A and n_B moles of A and B be modeled as

$$\Delta S_{mix} = -n_A R \cdot \log_e \left(\frac{n_A}{n_A + n_B} \right) - n_B R \cdot \log_e \left(\frac{n_B}{n_A + n_B} \right)$$

$$= -n_A R \cdot \log_e (X_A) - n_B R \cdot \log_e (X_B) \tag{7.4}$$

Let Equation (7.4) approximate entropic effects over and above those due to heat exchanges where ideal behavior is assumed for A and B. Let the entropy surrounding microscopic degrees of freedom, such as electronic, rotational, and vibrational, be encapsulated in the molar quantities S_A^o and S_B^o.

An example illustrates the salient points. Let the molar enthalpies and entropies of A and B be as follows:

$$H_A^o = 8000 \text{ joules / mole}, \; S_A^o = 5.00 \text{ joules / mole} \cdot Kelvin \tag{7.5A}$$

$$H_B^o = 6000 \text{ joules / mole}, \; S_B^o = 6.00 \text{ joules / mole} \cdot Kelvin \tag{7.5B}$$

Let the Figure 7.1 container commence with 1.00 mole of pure A at volume 1.00 meter³ and temperature 298 K. Then if x moles of B are formed locally, the entropy is modified eventually in two places—system and surroundings—as described by Equations (7.3) and (7.4), and ΔS_{AB}^o. The change for the surroundings is:

$$\Delta S_{surroundings} = \frac{-Q_{rec}}{T} = \frac{-(xH_B^o - xH_A^o)}{298 \text{ K}} = \frac{(xH_A^o - xH_B^o)}{298 \text{ K}} \tag{7.6}$$

while ΔS_{AB}^o equates with $xS_B^o - xS_A^o$. Figure 7.2 accordingly shows the dependence of each contribution to the entropy change; the total is included as a function of moles of B formed.

Several points follow. The reaction under consideration is exothermic because the molar enthalpy of A exceeds that of B; the A population loses heat-generating potential, so to speak, when members convert to B. In turn, the greatest entropy change for the heat bath would occur if all the A converted to B. The mixing ensures otherwise, however, by the maximum of ΔS_{mix} at $x = 0.500$ moles. The latter forces the total entropy to express a maximum somewhere between $x = 0.500$ and 1.00 mole, in this case, $x \approx 0.714$ *moles*. It is the extremum or apex of ΔS_{total} that ultimately dictates the equilibrium. As A switches to B, the sum of entropy changes increases until it can increase no more, regardless of mechanism. The extremum nature of ΔS_{total} retards the total conversion of A to B however skilled or wishful the chemist may be, for example, if B is a marketable commodity. At the same time, thermal environments are never devoid of fluctuations. Thus, in an equilibrium sample of A and B, accidental increases in one compound at the expense of the other switch on forces that push all parties back toward the maximum entropy state. It is not a coincidence that this behavior mirrors that of the helium–neon mixtures discussed in Chapter 3.

The apex of the total entropy change also determines the maximum available work—the free energy available from the system:

$$|\Delta G| = |-T \cdot \Delta S_{total}| \approx |-298 \text{ K} \cdot 10.48 \text{ joules / K}| \approx 3123 \text{ joules} \tag{7.7}$$

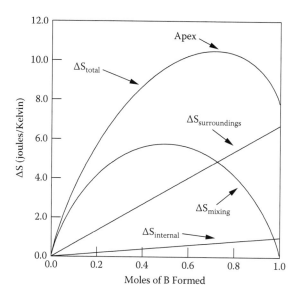

FIGURE 7.2 Entropy changes due to chemical reactions. The apex of ΔS_{total} locates the equilibrium condition.

There is really no limit to the possible initial conditions. If the system had featured 2.00 moles of pure A, then twice the work would have become available. If the chemist initiates matters by injecting (separately) 0.30 moles of A and 0.70 moles of B, then almost zero free energy would be offered. Note that the reaction switches two ways. If the initial conditions corresponded to A and B samples of 0.10 and 0.90 moles, respectively, then free energy would be obtained at the expense of the B population. For equilibrium conditions to exist and maintain, the reaction must be able to operate in forward and reverse directions. It is for this reason that the terms *reactant* and *product* are spoken largely for convenience. The reality is that B is the product of reactant A; A is the product of reactant B.

The role of the upper reservoir in Figure 7.1 is to intercept the available free energy somehow and relay it to where needed. If the reservoir-system coupling is faulty or compromised, then the available work is squandered. The laws of thermodynamics do not offer advice on how to reap the work of chemical reactions. The lessons are best taught by oxidation-reduction venues such as batteries and biological systems.

There is a second way to view the chemistry as is the subject of Figure 7.3. Shown is an A population distributed unevenly inside the container. The void near the center is where the local pressure is zero. Over this region, the chemical potential of A is:

$$\mu_A(T,p) = \mu_A^o(T) + RT \cdot \log_e(p_A)$$

$$= \mu_A^o(T) + RT \cdot \log_e(0)$$

$$= -\infty \qquad\qquad (7.8)$$

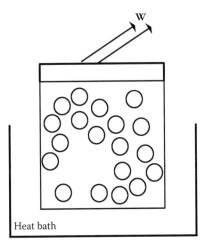

FIGURE 7.3   An alternate way to view chemically active systems. The exaggerated void represents a region where the chemical potential of A is negative infinity. At all points of the container, the chemical potential for B is negative infinity. The equilibration process is one that eases the potential differences.

Although the free-volume region has been exaggerated, the point should be clear, namely, that fluctuations cause p and μ to stray locally from their average values. The system responds by redistributing molecules and smoothing the gradients along the way. This characteristic brings additional effects when the molecules are chemically active.

A system of pure A is one where B exerts zero pressure and assumes the lowest possible chemical potential:

$$\mu_B(T,p) = \mu_B^o(T) + RT \cdot \log_e(p_B)$$

$$= \mu_B^o(T) + RT \cdot \log_e(0)$$

$$= -\infty \tag{7.9}$$

Under such conditions, the chemical potential of A is infinitely greater. The most direct way to ease the chemical gradient is for B molecules to be born at the expense of A; μ_B must increase while μ_A moves in the opposite direction.

Because of the second viewpoint, the lessons of Figure 7.2 can be extended. Let the standard chemical potentials of A and B be as follows:

$$\mu_A^o = 12,265 \text{ joules / mole} \qquad \mu_B^o = 10,000 \text{ joules / mole} \tag{7.10}$$

Let the initial conditions be identical to those responsible for Figure 7.2: $n_A = 1.00$ mole, $n_B = 0.00$ mole, $V = 1.00$ meter^3, and $T = 298$ K. The starting pressures and chemical potentials become

$$p_A = \frac{n_A RT}{V} = \frac{1.00 \text{ mole} \times 8.31 \dfrac{\text{joules}}{\text{mole} \cdot \text{K}} \times 298 \text{ K}}{1.00 \text{ meter}^3} \approx 2480 \text{ pascals} \quad (7.11)$$

$$\mu_A(T, p) = \mu_A^o(T) + RT \cdot \log_e(p_A) = 12,265 \text{ joules / mole}$$

$$+ 8.31 \frac{\text{joules}}{\text{mole} \cdot \text{K}} \times 298 \text{ K} \times \log_e(2480)$$

$$\approx 31,620 \text{ joules / mole} \quad (7.12)$$

$$p_B = \frac{n_B RT}{V} = \frac{0.00 \text{ mole} \times 8.31 \dfrac{\text{joules}}{\text{mole} \cdot \text{K}} \times 298 \text{ K}}{1.00 \text{ meter}^3} = 0.00 \text{ pascals} \quad (7.13)$$

$$\mu_B(T, p) = \mu_B^o(T) + RT \cdot \log_e(p_B) = 10,000 \text{ joules / mole}$$

$$+ 8.31 \frac{\text{joules}}{\text{mole} \cdot \text{K}} \times 298 \text{ K} \times \log_e(0.00)$$

$$= -\infty \quad (7.14)$$

If x moles of B are formed anywhere in the system, then the chemical potentials of both parties adjust accordingly:

$$\mu_A(T, p) = \mu_A^o(T) + RT \cdot \log_e\left(\frac{(1-x) \cdot RT}{V}\right) \quad (7.15)$$

$$\mu_B(T, p) = \mu_B^o(T) + RT \cdot \log_e\left(\frac{xRT}{V}\right) \quad (7.16)$$

Plots of the potentials appear in Figure 7.4. When A converts to B, the chemical potentials travel in opposite directions—one falls while the other rises—only to intersect at a single point. This intersection identifies the chemical equilibrium condition on equal footing with the apex of ΔS_{total}. This second viewpoint is complementary and emphasizes the equilibrium to be devoid of persistent and sharp chemical gradients. Thermal fluctuations are ever present. Yet any and all strays from maximum entropy only generate μ disparities, which steer the system back again.

A vital quantity is unique to the point of intersection. Since $\mu_A = \mu_B$ at maximum entropy, it follows that

$$\mu_A^o(T) + RT \cdot \log_e(p_A) = \mu_B^o(T) + RT \cdot \log_e(p_B) \quad (7.17)$$

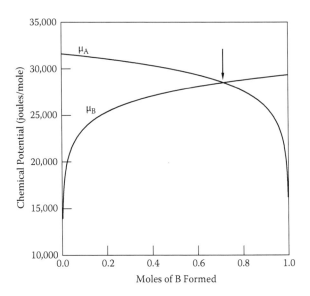

FIGURE 7.4 Chemical potentials of reactive molecules. When A switches to B, the chemical potentials move in opposite directions and intersect at a single point. The intersection marks the equilibrium condition in a manner complementary to ΔS_{total}.

Equation (7.17) can be rearranged to give:

$$\frac{\mu_A^o(T) - \mu_B^o(T)}{RT} = \log_e\left(\frac{p_B}{p_A}\right) \tag{7.18}$$

hence

$$\exp\left[\frac{\mu_A^o(T) - \mu_B^o(T)}{RT}\right] = \frac{p_B}{p_A} = K_p(T) \tag{7.19}$$

K_p is referred to as the mass action constant or simply the equilibrium constant. The terminology is misleading since Equation (7.19) shows K_p to be not a constant at all but rather an exponential function of temperature. Ideally, K_p does not depend on the initial amounts of A and B or any container properties—texture, shape, and so forth. In real-life cases, usually the best that the chemist can record is some value $K_p^{observed}$ due to the nonideality of reactants, products, and solvent. For the simple examples of Figures 7.2 and 7.4, the chemist would identify the respective equilibrium amounts of n_B and n_A to be 0.714 and $(1 - 0.714) = 0.286$ moles. Then

$$K_p = \frac{\dfrac{n_B RT}{V}}{\dfrac{n_A RT}{V}} = \frac{0.714}{0.286} \approx 2.50 \tag{7.20}$$

The chemist has a handle on K_p by his or her ability to measure concentrations and partial pressures. The fringe benefits include data regarding the difference between reactant and product chemical potentials. The qualitative lesson of Equation (7.19) is no less important. The greater the potential difference, the more skewed the reactant and product amounts at equilibrium.

The ideal dependence of K_p on temperature is not happenstance. The Gibbs–Helmholtz equation relates free energy G and enthalpy H as follows:

$$\left[\frac{\partial\left(\frac{G}{T}\right)}{\partial T}\right]_{p,n} = \frac{-H}{T^2} \tag{7.21}$$

In turn, the standard potentials—molar free energies—of A and B connect with the molar enthalpies as

$$\left[\frac{\partial\left(\frac{\mu_A^o}{T}\right)}{\partial T}\right]_{p,n} = \frac{-H_A^o}{T^2} \tag{7.22A}$$

$$\left[\frac{\partial\left(\frac{\mu_B^o}{T}\right)}{\partial T}\right]_{p,n} = \frac{-H_B^o}{T^2} \tag{7.22B}$$

These equations can be combined to give:

$$\left[\frac{\partial\left(\frac{\mu_A^o - \mu_B^o}{T}\right)}{\partial T}\right]_{p,n} = \frac{-H_A^o + H_B^o}{T^2} \tag{7.23}$$

Yet, Equations (7.18) and (7.19) already establish the relation between the chemical potential differences and temperature, namely,

$$\frac{\mu_A^o(T) - \mu_B^o(T)}{T} = R \cdot \log_e\left(\frac{p_B}{p_A}\right) \tag{7.24}$$

$$= R \cdot \log_e K_p(T)$$

One combines Equations (7.23) and (7.24) and arrives at the equality:

$$\left[\frac{\partial(R \cdot \log_e K_p(T))}{\partial T}\right]_{p,n} = \frac{-H_A^o + H_B^o}{T^2} \tag{7.25}$$

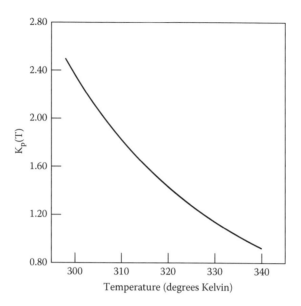

FIGURE 7.5   The equilibrium constant as a function of temperature. The dependence is exponential. Data pertain to A and B discussed in text.

Equation (7.25) is referred to as the van't Hoff equation. It asserts that the dependence of K_p on temperature is tied to the enthalpies of A and B. The qualitative aspects should not be overlooked. If $H_B^o > H_A^o$, K_p necessarily increases with temperature and vice versa. In highly atypical cases, $H_B^o = H_A^o$ and K_p is unaffected by temperature changes.

The ideal quantitative behavior is illustrated in Figure 7.5. For finite changes in the temperature ΔT, one has from Equation (7.25):

$$\left[\frac{\partial (R \cdot \log_e K_p(T))}{\partial T} \right]_{p,n} \approx \left[\frac{R \cdot \log_e K_p(T+\Delta T) - R \cdot \log_e K_p(T)}{\Delta T} \right]_{p,n} \quad (7.26)$$

Combining Equations (7.25) and (7.26), one has

$$R \cdot \log_e K_p(T+\Delta T) \approx \frac{\Delta T \cdot (-H_A^o + H_B^o)}{T^2} + R \cdot \log_e K_p(T) \quad (7.27)$$

and subsequently

$$K_p(T+\Delta T) \approx \exp\left[\frac{\Delta T \cdot (-H_A^o + H_B^o)}{RT^2} + \log_e K_p(T) \right] \quad (7.28)$$

Equation (7.28) enables the construction of $K_p(T)$ plots over small to modest temperature ranges. Figure 7.5 augments the lessons of Figure 7.2 and Figure 7.4 by showing a typical exponential dependence of K_p on temperature. The crucial matter to observe is that small changes in temperature effect substantive responses by K_p. Note that Figure 7.5 has been constructed with the assumption that the molar enthalpies of A and B are more or less constant. This assumption is generally valid over a few tens of degrees and modest deviations in the pressure from standard states.

The discussion has been limited to a case with trivial stoichiometry—molecule A begets a single B and vice versa. What if the number of molecules is not conserved? Consider, for example:

$$A + 2B \rightleftharpoons C$$

where three charge packages combine to produce one in a left-to-right transition; one converts to three in the opposite direction.

The lessons of Figure 7.2 are not impacted qualitatively by stoichiometry. The equilibrium is still governed by the maximum in ΔS_{total}. There are details to point out, however, regarding the chemical potentials and mechanisms by which the entropy is tuned. These details make for complications quickly, even when the molecules are viewed as ideal gases.

First, if the chemist injects A, B, and C individually into a container in arbitrary amounts, there is an eventual entropic impact due to mixing. This can be modeled by an extension of Equation (7.4):

$$\Delta S_{mix} = -n_A R \cdot \log_e \left(\frac{n_A}{n_A + n_B + n_C} \right) - n_B R \cdot \log_e \left(\frac{n_B}{n_A + n_B + n_C} \right)$$

$$- n_C R \cdot \log_e \left(\frac{n_C}{n_A + n_B + n_C} \right) \tag{7.29}$$

$$= -n_A R \cdot \log_e (X_A) - n_B R \cdot \log_e (X_B) - n_C R \cdot \log_e (X_C)$$

Again, the chemist would have to spend work to undo all or part of the scrambling of the molecules. If instead, the system is left alone, chemical reactions will impact the composition and entropy ejected to or imported from the surroundings. If the chemist injects only A and B separately, and x moles of C are formed at the interface, then the mixing effects can be modeled via:

$$\Delta S_{mix} = -(n_A - x)R \cdot \log_e \left(\frac{n_A - x}{n_A + n_B - 2x} \right) - (n_B - 2x)R \cdot \log_e \left(\frac{n_B - 2x}{n_A + n_B - 2x} \right)$$

$$- xR \cdot \log_e \left(\frac{x}{n_A + n_B - 2x} \right) \tag{7.30}$$

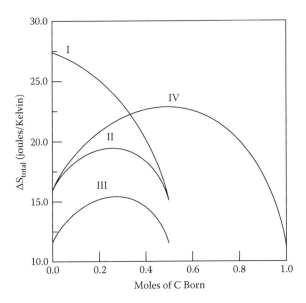

FIGURE 7.6   Total entropy changes for chemically active molecules. Data pertain to A, B, and C molecules and scenarios discussed in the text.

while the entropy of the heat bath alters as follows:

$$\Delta S_{surroundings} = \frac{-Q_{rec}}{T} = \frac{-(xH_C^o - xH_A^o - 2xH_B^o)}{298 \text{ K}} = \frac{+(xH_A^o + 2xH_B^o - xH_C^o)}{298 \text{ K}} \quad (7.31)$$

Equation (7.31) basically parallels Equation (7.6), taking into account the stoichiometry details.

One then notes a consequence of entropy being an extensive quantity. Although conversions of A and 2B to C can enhance the mixing possibilities, they detract from the total moles of particles. There is free energy available from reactant → product conversions. Yet the spontaneous formation of C means less work available for other potentially useful tasks such as container expansion. It is the general rule that entropy scales with the number of particles times a factor of R: $S \propto nR$. The precise scaling hinges on the electronic structure of the reactants and products, in addition to the temperature and container volume. In the simplest approaches, these effects are packaged as $\Delta S_{ABC}^o = xS_C^o - xS_A^o - 2xS_B^o$, where x is the moles of C formed.

One considers the impact of stoichiometry under idealized circumstances. Figure 7.6 charts four (of infinite possible) scenarios based on the following initial conditions, all species injected separately by the chemist into the container:

Scenario I: $n_A = n_B = n_C = 1.00$ mole
Scenario II: $n_A = 2.00$ mole; $n_B = 1.00$ mole; $n_C = 0.00$ mole
Scenario III: $n_A = n_B = 1.00$ mole; $n_C = 0.00$ mole
Scenario IV: $n_A = 1.00$ mole; $n_B = 2.00$ mole; $n_C = 0.00$ mole

Let the molar enthalpies, entropies, and free energies of A and B be as before (Equations 7.5 and 7.10). Let

$$H_C^o = 14,000 \text{ joules / mole}, \quad S_C^o = 8.00 \text{ joules / mole} \cdot \text{Kelvin} \quad (7.32)$$

$$\mu_C^o = 71,008 \text{ joules / mole} \quad (7.33)$$

and the container be (as before) 1.00 meter^3 in volume at constant temperature 298 K.

Figure 7.6 shows the dependence of the total entropy change for each case. Each curve is obtained by summing the entropy changes due to mixing and heat exchanges with the surroundings plus ΔS_{ABC}^o. The lessons are several, beginning with the single maximum demonstrated by each curve. Each maximum hinges on the interplay of ΔS_{mix}, $\Delta S_{surroundings}$, and ΔS_{ABC}^o, in addition to choices made by the chemist. A single maximum, in each case, offers a foundation for stability, regardless of initial conditions. For every case, fluctuations will push the molecules on either side of the maximum entropy state yet will always turn on forces for restoration.

The chemist's choices are always critical. Scenarios II and III show only modest increases in ΔS_{total} with the birth of C. Modest values of ΔS_{total} predicate modest free energy available. Thus, if the chemist needs work to be performed, clearly he or she should seek conditions other than scenarios II and III; I and IV are the most appealing in Figure 7.6. The largest entropy changes are offered in scenario I; these transpire at the expense of the C population. By comparing the scenarios, one gathers that the production of C is greatest when reactants A and B are combined in stoichiometric amounts, that is, in a 1:2 mole ratio.

Equilibrium conditions hold when there is no additional free energy for the system to lose:

$$G_{total} = G_{react} + G_{products} = minimum \ possible \ value \quad (7.34)$$

Thus, when an A, B, C sample is at chemical equilibrium, and a fluctuation causes δx moles of C to be born at, say, the reactants' expense, the free energy responds accordingly:

$$\delta G_{total} = -\delta x \cdot \mu_A + -\delta x \cdot 2 \cdot \mu_B + \delta x \cdot \mu_C \quad (7.35)$$

But this adjustment tends to zero by the stability of the maximum entropy state. By moving the Equation (7.35) terms to one side and factoring δx, one obtains:

$$0 = +\delta x \cdot (\mu_C - \mu_A - 2\mu_B) \quad (7.36)$$

Fluctuations are ever present and render infinite possible values for δx. Equation (7.36) can hold only if, for equilibrium conditions, the terms in parentheses equate to zero:

$$\mu_C - \mu_A - 2\mu_B = 0 \quad (7.37)$$

This conveys that the chemical potentials are related to each other (at equilibrium) in a manner governed by the reaction stoichiometry:

$$1 \cdot \mu_A + 2 \cdot \mu_B = 1 \cdot \mu_C \tag{7.38}$$

Taking A, B, and C to behave as ideal gases, it is straightforward to substitute for the chemical potentials (cf. Equations 7.17 through 7.19) and rearrange terms so as to obtain the mass action constant under ideal conditions:

$$K_p(T) = \exp\left[\frac{\mu_A^o(T) + 2\mu_B^o(T) - \mu_C^o(T)}{RT}\right] = \frac{p_C}{p_A \cdot p_B^2} \tag{7.39}$$

As before, K_p is an exponential function of temperature that is determined by differences in chemical potentials. And usually the chemist must settle for $K_p^{observed}$ in an experiment, due to nonideality of the active parties and solvent. More important, there is one inconsistency to note. The argument of the exponential, and thus K_p, is dimensionless. Yet when K_p is expressed in terms of equilibrium pressures raised to powers set by the stoichiometry, obtained are (in this case) dimensions of pascals^{-2}. This mismatch is an unfortunate side effect of writing chemical potentials with a solitary pressure in the logarithm argument. The more exacting, if cumbersome, way to express the potential for molecule A is:

$$\mu_A(T, p) = \mu_A^o(T) + RT \cdot \log_e\left(\frac{p_A}{1.00 \text{ pascal}}\right) \tag{7.40}$$

with like attention to B and C. The extra labor does not really offer additional insights, but it does lead to a consistent, dimensionless version of the mass action constant:

$$K_p(T) = \exp\left[\frac{\mu_A^o(T) + 2\mu_B^o(T) - \mu_C^o(T)}{RT}\right] = \frac{\left(\dfrac{p_C}{1.00 \text{ pascal}}\right)}{\left(\dfrac{p_A}{1.00 \text{ pascal}}\right) \cdot \left(\dfrac{p_B}{1.00 \text{ pascal}}\right)^2} \tag{7.41}$$

Plots of the chemical potentials are presented in Figure 7.7 as a function of the moles of C formed. The conditions reflect scenario IV of the preceding figure. The sum of the A and B chemical potentials weighted by stoichiometric coefficients is shown along with the potential of C. The lessons are clear. The C potential starts at its minimum, impossible-to-plot value—negative infinity. As the system proceeds toward chemical equilibrium, $\mu_A + 2\mu_B$ and μ_C move in opposite directions and eventually intersect. It is straightforward to construct links between K_p and the molar enthalpies, in addition to relations appropriate to other stoichiometric conditions. Where and when there is

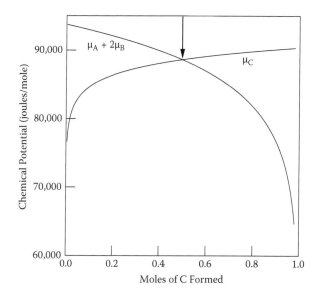

FIGURE 7.7   Chemical potentials of reactive molecules. Data pertain to molecules A, B, and C discussed in text. The conditions correspond to IV of the preceding figure.

under- or overshoot, gradients in the chemical potentials direct the system back toward the maximum entropy state, the point of $\mu_A + 2\mu_B = \mu_C$ intersection. This revisits an important point made in Chapter 3, namely, that the equilibrium condition is not singular but rather comprises multiple states in the vicinity of maximum entropy. Regardless of reaction mechanism and stoichiometry, when a system demonstrates the maximum possible entropy, a minimum number of variables suffice to describe it. Whenever the system strays from maximum entropy, more information is needed by the chemist to detail the conditions.

7.2   CHEMICAL REACTIONS AND INFORMATION

Chemical reactions do not substantially alter information in the statistical sense at the macroscopic scale. If sparse information is allied with p, V, and T when A owns the container, the same applies after equilibrium has been established with B. There is little uncertainty encountered by the chemist prior to measuring any number of state quantities. The microscopic scale is another story, however.

Molecules are electric charge packages that communicate by thermal collisions. A container of pure A offers only one type of message—ignoring structure considerations for the moment. For a given A molecule, there is no uncertainty imposed on the next binary collision—it will assuredly involve another A. Matters are different when B shares the container, either injected by the chemist or born via the demise of A. For an equilibrium mixture, the collisions fall into three categories: AA, AB, and BB. Any effects of higher order contacts (e.g., ternary collisions), molecular speeds, and trajectories fall outside the scope of the discussion.

An electrical contact happens because two molecules move accidentally, and more or less independently, toward the same territory. As the motion is thermally powered—and thus random—the likelihood of a party vying for a particular spatial region is proportional to its mole fraction. Thus for an A,B mixture, the probability of each particular type of binary collision is proportional to the products of mole fractions, namely,

$$prob(A,A) \propto X_A \cdot X_A \tag{7.42A}$$

$$prob(A,B) = prob(B,A) \propto X_A \cdot X_B \tag{7.42B}$$

$$prob(B,B) \propto X_B \cdot X_B \tag{7.42C}$$

whereupon

$$prob(A,A) = \frac{X_A \cdot X_A}{X_A \cdot X_A + X_A \cdot X_B + X_B \cdot X_B} \tag{7.43A}$$

$$prob(A,B) = \frac{X_A \cdot X_B}{X_A \cdot X_A + X_A \cdot X_B + X_B \cdot X_B} \tag{7.43B}$$

$$prob(B,B) = \frac{X_B \cdot X_B}{X_A \cdot X_A + X_A \cdot X_B + X_B \cdot X_B} \tag{7.43C}$$

Figure 7.2 illustrated the entropy properties of the A,B chemistry. The initial conditions featured only one type of message event, namely, AA, while the equilibrium and intermediate states afforded three. Figure 7.5 showed that the mass action constant changes with temperature. It is interesting to examine the Shannon information $I_{A,B}$ allied with the message events and their link to K_p. From Equation (7.20), we have

$$K_p = \frac{\dfrac{n_B RT}{V}}{\dfrac{n_A RT}{V}} = \frac{\dfrac{n_B}{n_A + n_B}}{\dfrac{n_A}{n_A + n_B}} = \frac{X_B}{X_A} \tag{7.44}$$

As a consequence

$$X_B = K_p \cdot X_A \tag{7.45}$$

and the message event probabilities can be rewritten in terms of K_p:

$$prob(A, A) = \frac{X_A^2}{X_A^2 + X_A^2 \cdot K_p + X_A^2 \cdot K_p^2}$$

$$= \frac{1}{1 + K_p + K_p^2} \tag{7.46A}$$

$$prob(A, B) = \frac{X_A^2 \cdot K_p}{X_A^2 + X_A^2 \cdot K_p + X_A^2 \cdot K_p^2}$$

$$= \frac{K_p}{1 + K_p + K_p^2} \tag{7.46B}$$

$$prob(B, B) = \frac{X_A^2 \cdot K_p^2}{X_A^2 + X_A^2 \cdot K_p + X_A^2 \cdot K_p^2}$$

$$= \frac{K_p^2}{1 + K_p + K_p^2} \tag{7.46C}$$

The Shannon information is obtained in the usual way from summing weighted logarithmic terms. There are three different message events and thus terms in the summation. One has

$$I_{A,B} = -\sum_{i=1}^{3} prob(i) \cdot \log_2(prob(i)) \tag{7.47}$$

or, more specifically,

$$I_{A,B} = \frac{-1}{\log_e(2)} \times \left[prob(A, A) \cdot \log_e prob(A, A) + prob(A, B) \cdot \log_e prob(A, B) \right.$$
$$\left. + prob(B, B) \cdot \log_e prob(B, B) \right] \tag{7.48}$$

where the probability terms are specified by Equation (7.46A) through (7.46C). Figure 7.8 shows $I_{A,B}$ versus K_p. A single maximum is observed where nature extends no bias toward the reactant or product. Thus, the information maximum applies to the conditions whereby K_p equals 1.00. Figure 7.5 tells us that such conditions are found when the temperature is set to approximately 336 K. Note the obvious. If A had been chemically inert, then the number of message types would have remained at 1; no information would have been purchased unless solvent molecules were added

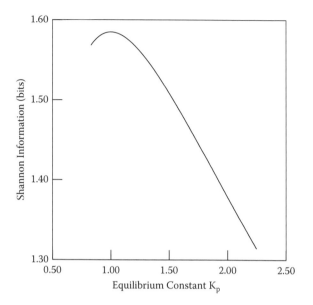

Shannon Information (bits)

Equilibrium Constant K_p

FIGURE 7.8 Information versus equilibrium constant. A single maximum is observed. Data pertain to the A,B reactions discussed in the text.

to the system. By virtue of A's capacity to convert to B, information is born spontaneously at a cost of free energy. Chemical activity is a means for nature to purchase and utilize new information.

It is likewise important to consider the mutual information. A and B have been viewed as ideal gases. This means a lack of preference shown by one party toward colliding with another. The consequence is that a binary collision at one locale Y of the system offers nothing about another locale Z. In turn,

$$prob\left[(A, A)_Y,(A, A)_Z\right] = prob(A, A)_Y \times prob(A, A)_Z \qquad (7.49A)$$

$$prob\left[(A, A)_Y,(A, B)_Z\right] = prob(A, A)_Y \times prob(A, B)_Z \qquad (7.49B)$$

$$prob\left[(A, A)_Y,(B, B)_Z\right] = prob(A, A)_Y \times prob(B, B)_Z \qquad (7.49C)$$

$$prob\left[(A, B)_Y,(B, B)_Z\right] = prob(A, B)_Y \times prob(B, B)_Z \qquad (7.49D)$$

$$prob\left[(B, B)_Y,(B, B)_Z\right] = prob(B, B)_Y \times prob(B, B)_Z \qquad (7.49E)$$

$$prob\left[(A, B)_Y,(A, B)_Z\right] = prob(A, B)_Y \times prob(A, B)_Z \qquad (7.49F)$$

There are six terms to consider regarding the mutual information. However, all have logarithm arguments of 1, for example:

$$prob\left[(A, A)_Y, (A, B)_Z\right] \cdot \log_2 \left\{ \frac{prob\left[(A, A)_Y, (A, B)_Z\right]}{prob(A, A)_Y \times prob(A, B)_Z} \right\} \qquad (7.50)$$

The result is that *MI* terms are uniformly zero for ideal gas behavior. It is only when interactions between the molecules are turned on that *MI* will exceed zero. The interactions impose clustering and excluded volume effects, as in the van der Waals model. These confer a statistical spatial structure in the system.

7.3   REACTIONS, INFORMATION, AND MOLECULAR STRUCTURE

There is additional information to address at the Angstrom scale. It was stated that several properties are conserved in a reaction: energy, mass, charge, and atom identity. Electronic structure is not one of them, fortunately. A reactant and product accordingly express different facts and data information. We look briefly at how information changes in the statistical sense.

The following are reaction examples of the A = B variety:

Reaction 1 describes a diketone-enol tautomerization, whereas 2 and 3 are Cope rearrangements discussed in second-year chemistry courses and beyond. It is important that information in the statistical sense differs for the left and right sides because the atom–bond bond networks predicate different collision sequence possibilities.

One appeals to the Chapter 6 methods that treat compounds as Brownian computers. An organic molecule is composed of atom–bond–atom units C-H, C=C, and so on, and communicates via collisions. A nearest-neighbor random (thermal) walk over each formula graph identifies the possible electronic messages, for example, for 2,4-pentanedione

...(C-C)(C-C)(C-H)(C-C)(C=O)(C-C)(C=O)(C-C)(C-H)(C-H)(C-C)...

as opposed to

...(C-H)(C-H)(C-C)(C=C)(C-O)(O-H)(C-O)(C-C)(C-H)(C-C)(C-H)...

for the enol tautomer.

It was shown in Chapter 6 that the contrasts among molecules are most acute concerning mutual information and energy dispersion. Following the same approach, reduced descriptors for the states can be constructed using ethane data for a baseline, for example:

$$\hat{\xi}_{MI}^{2,4-pdione} = \frac{\xi_{MI}^{2,4-pdione}}{\xi_{MI}^{ethane}} \tag{7.51}$$

$$\hat{\xi}_{\sigma D}^{2,4-pdione} = \frac{\xi_{\sigma D}^{2,4-pdione}}{\xi_{\sigma D}^{ethane}} \tag{7.52}$$

This analysis of the three reactions leads to Figure 7.9. The state points for reactant and products (left- and right-side compounds) have been distinguished by filled and open symbols, respectively. The numbers correspond to the labels attached to the equilibration symbols. Error bars have been included as established by the $\hat{\xi}$ uncertainties in linear regression analyses.

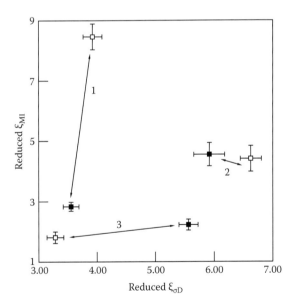

FIGURE 7.9 Reduced descriptors for chemical reactants and products. Plotted are $\hat{\xi}_{MI}$ versus $\hat{\xi}_{\sigma D}$ scaled via ethane data. 1, 2, and 3 refer to the unimolecular reactions discussed in the text. Filled symbols denote the state points for reactants; open symbols locate the state points of products.

The results are striking. In the diketone $\rightarrow$ enol transformation, the mutual information of the collision sequence, and thus $\hat{\xi}_{MI}$, is enhanced by almost a factor of 3. The correlations among the electronic messages are multiplied significantly in spite of the formula invariance (i.e., $C_5H_8O_2 \rightarrow C_5H_8O_2$) and skeletal similarities. The changes derive from three message units of the diketone, C-H, C=O, and C-C, increasing by two, C=C and O-H, as the enol is born. Reactions switch in two directions. Thus, the message correlations are diminished significantly when an enol converts to the diketone.

In the Cope rearrangements, the changes are minor in $\hat{\xi}_{MI}$ but substantive in $\hat{\xi}_{\sigma_p}$. This is in spite of the left and right sides hosting identical ABA (and thus electronic message) constituents. The dispersion energy is sensitive to changes in the collision sequences allowed by the covalent bond networks. $\hat{\xi}_{\sigma_p}$ can be augmented in the left-to-right direction of reaction 2 and diminished as in 3. It is important that for all reactions the transformations are not smooth and continuous as in pressure and volume tuning at the macroscopic scale. A chemical reaction rather marks a jump relocation of the microscopic state point.

Reactants and products are not restricted to single molecules. The following describe combinations in the forward (left to right) direction, dissociations in reverse.

Reaction 4 is the famous Diels–Alder, whereas 5 was developed decades later by Longley, Emerson, and Blardinelli [1]. Information analysis of the collision-based message spaces leads to Figure 7.10. The same format is used as in the previous figure. Dotted lines are drawn to indicate the coupling of reactant states.

The feature to note is that the state point of the product compound is not placed merely by adding the coordinates of the reactants. By the same token, the product coordinate is not an average of the parents. This means that information in the statistical sense and dispersion energy are not conserved in typical reactions—there is a nonzero distance between the state points of left- and right-side compounds. Further, the jump relocations of the points do not simply land at mean (center of mass) values.

There is a final lesson to note by revisiting reaction 1 concerning 2,4-pentanedione and its enol tautomer. If a chemist prepared an equilibrium sample of the left- and right-side molecules, he or she would obtain a system for which the following holds true:

$$\mu_{diketone}(T, p) = \mu_{enol}(T, p) \tag{7.53}$$

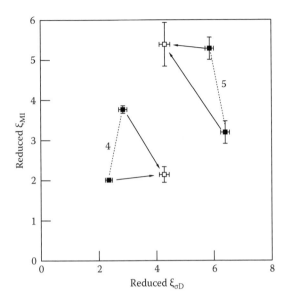

FIGURE 7.10 Reduced descriptors for reactants and products. Plotted are $\hat{\xi}_{MI}$ versus $\hat{\xi}_{\sigma_D}$ scaled via ethane data. 4 and 5 refer to bimolecular reactions discussed in the text. Filled symbols denote the state points for reactants; open symbols refer to products.

Perhaps overlooked is the large number of molecules with chemical potential infinitely lower than $\mu_{diketone}$ and μ_{enol}. For example, the following are structural isomers of 2,4-pentanedione and the enol tautomer:

Both the 1,3-diol of cyclopentene and the carboxylic acid derivative of cyclobutane have formula $C_5H_8O_2$, yet there is no accessible mechanism for establishing chemical equilibrium with the 2, 4-pentanedione or its enol tautomer. The chemical potentials of both molecules would be fixed at negative infinity in the system. There are additional structural isomers for which the equivalent statement can be made.

The lesson follows from Figure 7.11. Shown are the state points for the reaction 1 compounds along with the $\hat{\xi}_{MI}$, $\hat{\xi}_{\sigma_D}$ coordinates for the aforementioned diol of cyclopentene and the acid derivative of cyclobutane. Reaction 1 represents a jump relocation of the state point as shown. The process need not couple with, or move smoothly, through nearby states. Molecular information changes selectively, as programmed by the electronic structure. It is the purview of synthetic chemistry to establish the selection rules by theory and experiment, and to optimize the programming strategies.

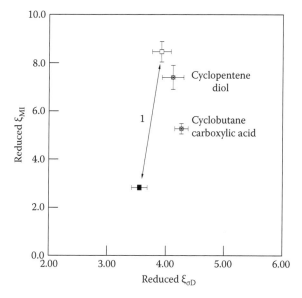

FIGURE 7.11 Reduced descriptors for tautomers and structural isomers of $C_5H_{10}O_2$. Plotted are $\hat{\xi}_{MI}$ versus $\hat{\xi}_{\sigma_D}$ scaled via ethane data. 1 refers to the tautomerization of 2,4-pentanedione. The two other state points in the figure derive from structural isomers.

The consequences of reactions relocating state points spontaneously and selectively are not trivial. If the chemist owns a bottle containing 2,4-pentanedione but needs the 1,3-diol of cyclopentene, a conversion program would be in order. Executing the program would entail a sequence of intermediate molecules, each with a unique electronic message space and state point in the $\hat{\xi}_{MI}$, $\hat{\xi}_{\sigma_D}$ plane. The chemist would devise a program for sequential transformations that minimizes the distance traveled across the plane. Alternatively, he or she would seek a starting material other than 2,4-pentanedione, which affords a shorter program that terminates in the desired state point.

The major points of this chapter are the following:

1. The sun is the ultimate source of atoms. Molecules, by and large, descend from other molecules. The transformations occur by chemical reactions, which increase the total entropy and yet modify information locally and selectively.

2. Reactions alter the electronic messages of a molecule. The modifications impact the information expressed via binary or higher order collisions. The changes are internal to the molecules as well. These can be characterized by formula diagrams and further modeled by Brownian computation.

3. At the microscopic level, every molecule offers a state point and width in planes such as $\hat{\xi}_{MI}$, $\hat{\xi}_{\sigma_D}$. A reaction relocates a state point in a jump-discontinuous way. The direction and extent of the jump are determined by the electronic information and correlations.

7.4 SOURCES AND FURTHER READING

Thermodynamic texts devote one or more chapters to chemical reactions. The texts by Desloge [2], Fermi [3], Lewis and Randall [4], Kirkwood and Oppenheim [5], and Klotz [6] have particularly impacted the author's thinking. Desloge and Fermi offer exceptional presentations of so-called van't Hoff boxes containing reactants and products. These devices offer an especially insightful perspective of chemical equilibrium. Regarding the microscopic scale, the author and students have applied the Brownian computer model to a variety of chemical reactions, including diketone-enol tautomerizations and Cope rearrangements [7]. These reactions are discussed at length in classic texts by Wheland [8] and le Noble [9]. Goodstein's first-chapter summary of thermodynamics offers penetrating insights regarding the entropy of ideal gases [10].

7.5 SUGGESTED EXERCISES

7.1 Molecule A reacts to form B and vice versa in the gas phase. A chemist finds K_p for the reaction to equal 1.20 at temperature 298 K. (a) What is the value of $\mu_A^o - \mu_B^o$ at 298 K? (b) The chemist adds 2.00 moles of B and 0.500 moles of A to a container of volume 1.50 meter^3 at 298 K. How many bits of Shannon information are allied with each binary collision at equilibrium?

7.2 A chemist finds that K_p for the gas phase reaction 2A + 3B = 0.50C + 3D equals 0.150 at temperature 298 K, the partial pressures having been measured in pascals. K_p is found to double when the temperature is raised to 303 K. (a) What is the value of $2\mu_A^o + 3\mu_B^o - 0.50\mu_C^o - 3\mu_D^o$ at 298 K? (b) What is the approximate value at 300 K?

7.3 There is more than one equation of the Gibbs–Helmholtz variety. (a) Show that

$$\left[\frac{\partial \left(\frac{A}{T} \right)}{\partial T} \right]_{V,n} = \frac{-U}{T^2}$$

(b) Show that

$$\left[\frac{\partial \left(\frac{\Phi}{\mu} \right)}{\partial \mu} \right]_{V,S} = \frac{-U}{\mu^2}$$

where Φ was encountered in Chapter 3 as a Legendre transform of U.

7.4 Consider A that reacts to form B and vice versa in the gas phase with equilibrium constant K_p. Let $\mu_B^o - \mu_A^o$ equal the molar free energy change of an A,B-system coming to equilibrium at temperature T, given initial (standard) conditions of $p_A = p_B = 1.00$ pascal. (a) For arbitrarily selected initial conditions at T, show that

$$\mu_B - \mu_A = -RT \log_e (K_p) + RT \log_e \left(\frac{p_{B,initial}}{p_{A,initial}} \right).$$ (b) How should the

equation of part (a) be interpreted if either A or B is ever at zero pressure? Please discuss.

7.5 Chemical reactions can change the moles of molecules in a system by amount Δn. Show that for ideal gases

$$S(n + \Delta n) \approx S(n) \cdot \left[1 + \frac{\Delta n}{n} \right] - R\Delta n \cdot$$

7.6 The application of the mixing entropy in this chapter is approximate at best. (a) Discuss the conditions for the A = B chemistry that enhance the applicability. (b) Do likewise for the A + 2B = C reaction. (c) When can the mixing entropy be ignored for chemical equilibration processes? Please discuss.

7.7 Consider the case where molecule A collides with a structural isomer B and is able to convert to yet another B. A chemist finds $K_p = 1.20$ at 298 K. He or she adds 2.45 and 1.40 moles of A and B, respectively, to a 5.00 meter^3 container at 298 K. (a) Construct a plot of the bits of information per binary collision versus moles of A converted. (b) Construct a plot of the entropy of mixing versus moles of A converted. (c) Construct a plot of the bits of information per binary collision versus joules of work lost by the system.

7.8 A chemist studies the reaction A + 2B = C with $K_p = 0.125$ at temperature 298 K. (a) In one experiment, the chemist adds 1.00 mole each of A, B, and C to a 5.00 meter^3 container at 298 K. Construct a plot of the bits of information per binary collision as a function of moles of C converted. How many bits correspond to the equilibrium state? (b) Consider the effects of higher order collisions: construct a plot of the bits of information per ternary collision as a function of moles of C converted. How many bits correspond to the equilibrium state? (c) How do the information amounts compare for binary and ternary collisions? Please discuss.

7.9 2-4-pentanedione poses four tautomeric forms:

Use the Chapter 6 methods to locate the state point in the $\hat{\xi}_{MI}$, $\hat{\xi}_{\sigma_D}$ plane for each molecule. For which molecule is the state point closest to that of 2-4-pentanedione? Farthest? Please discuss.

7.10 A reacts to form B and vice versa; B reacts to form C and vice versa. One version of this is as follows:

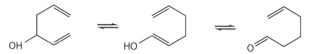

Use the Chapter 6 methods to locate the state point in the $\hat{\xi}_{MI}$, $\hat{\xi}_{\sigma_D}$ plane for each molecule. Which pair of points is closest in the state space? Please discuss.

7.11 A chemist imagines a reaction A = B where $\mu_A^o = \mu_B^o$, $H_A^o = H_B^o$, and $S_A^o = S_B^o$. (a) What type of reaction is being contemplated? (b) What is the value of K_p regardless of temperature? (c) The chemist adds 0.385 and 1.40 moles of A and B, respectively, to a 5.00 meter^3 container at 298 K. Construct a plot of the bits of information per binary collision versus moles of A converted. (d) Construct a plot of the entropy of mixing versus moles of A converted.

REFERENCES

[1] Longley, R. I., Emerson, W. S., Blardinelli, A. J. 1954. 3,4-dihydro-2-methoxy-4-methyl-2H-pyran, *Organic Syntheses* 34, 29.
[2] Desloge, E. A. 1968. *Thermal Physics*, Holt, Rinehart, and Winston, New York.
[3] Fermi, E. 1956. *Thermodynamics*, Dover, New York.
[4] Lewis, G. N., Randall, M. 1923. *Thermodynamics and the Free Energy of Chemical Substances*, McGraw-Hill, New York.
[5] Kirkwood, J. G., Oppenheim, I. 1961. *Chemical Thermodynamics*, McGraw-Hill, New York.
[6] Klotz, I. 1964. *Introduction to Chemical Thermodynamics*, W. A. Benjamin, New York.
[7] Graham, D. J., Schulmerich, M. V. 2004. Information Content in Organic Molecules: Reaction Pathway Analysis via Brownian Processing, *J. Chem. Inf. Comput. Sci.* 44, 1612.
[8] Wheland, G. W. 1944. *The Theory of Resonance and Its Application to Organic Chemistry*, Wiley, New York.
[9] le Noble, W. J. 1974. *Highlights of Organic Chemistry*, Dekker, New York.
[10] Goodstein, D. L. 1985. *States of Matter*, chap. 1, Dover, New York.

8 Chemical Thermodynamics, Information, and Horizons

8.1 HORIZONS

When it comes to information and chemistry, there is no shortage of challenges. Molecules carry information and communicate via collisions. If the chemist thoroughly understood the messages in amino acid sequences, he or she would be able to design proteins from the ground up. Further, comprehension of axon and neuron information at the molecular level would provide new insights into neurological therapy. Further still, understanding how the information stored in a virus fluctuates over replication cycles would reveal new defense strategies. These are but three areas that are investigated internationally. The word *information* will be included in science press releases on and over the horizon.

This text has examined terrain far removed from blockbusters. The result has been a partial exploration of two mature domains and where they intersect. Thermodynamics and information theory have supplied decades of fundamentals and applications. Insights have arrived in both disciplines by their governing laws, mathematical structure, and idealized models. Heat engines do not conform to Carnot cycles. The states of a system are generally not 100% known in advance of an experiment. Molecules are not formula diagrams written on paper. Yet the laws, mathematics, and models surrounding engines, states, and molecules steer the investigator in the right direction. At the very least, they show what properties the chemist should become more curious about.

Where does one travel next at the intersection? There were corners notably untouched: third-law consequences, phase rule applications, and critical phenomena, to name three. These justifiably warrant full-scale treatments. By the third law of thermodynamics, the entropy and heat capacity of a crystalline system converge to zero in the limit of zero absolute temperature. According to Boltzmann:

$$S = k_B \cdot \log_e(W) \qquad (8.1)$$

where W is the number of possible states for a system. Clearly, W converges to unity as T moves toward zero. As witnessed on several occasions of the tour, systems that pose only one state present no information in the statistical sense. Horizons-wise, the properties of thermodynamic information at ultralow temperatures are well worth exploring.

Phase rule applications merit the same attention. According to Gibbs:

$$\phi + \rho = \kappa + 2 \tag{8.2}$$

where ϕ and κ are the number of phases and components, respectively, for a system at equilibrium. ρ represents the so-called freedom of a system—the number of intensive variables that need to be set by the chemist to control all the intensive quantities. For a system with $\phi = 2$ and $\kappa = 2$, the chemist must hold, say, pressure and temperature constant to maintain the equilibrium. The chemist is certainly capable of arbitrarily fixing a third intensive variable such as specific heat. But this would mean the demise of one of the phases. As seen during the tour, information connects with the control capacity of a system. The thermodynamic information surrounding the equilibrium between phases is a second horizon to explore.

Phase-rule effects lie not far away from phase transitions. The latter invite attention because of the multiplication issues surrounding information. Information begets information and sometimes a little can control a lot. In certain phase transitions, the information expressed by a seed material, liquid or crystalline, poses a blueprint for constructing a much larger version. The information stored in the seed is multiplied many times over in a high-fidelity fashion. *Many* does not equate with *infinite*; one would anticipate an upper bound for the number of multiplication cycles, depending on the thermal circumstances. This is a third horizon to sail toward, again not far removed from phase-rule effects.

Last, capital T received much attention during the tour via temperature, heat capacity, and entropy. Small t received virtually zero in the manner of time. This was understandable given the intended level of the text and the spotlight on equilibrium conditions. Recall that these rarefied conditions offer no information about the past or future. Thus, any incorporation of t with heat, work, and information links to nonequilibrium thermodynamics. This is a domain markedly different from that of the present text in principles, models, and applications. It emphasizes the second half of the word *thermodynamics*.

Clearly, t needs to be explored as the fourth horizon. It needs to be integrated with the core topics of fluctuations, state transformations, chemical message transmission, and registration without sacrificing the accessibility. Time is unusual as resources go because, as mentioned in Chapter 1, there is never really a source or recipient. Time wields impact to the same degree as work, heat, and information. No information is transmitted and registered, no energy is transferred, and no state is transformed without some elapse of t. Time acquires special meaning if a system deviates from the equilibrium condition, whether slightly or drastically. The pathway by which the system strives toward the maximum entropy hinges on time, both the amount elapsed and available. Time issues are not critical just to chemical thermodynamics; they impact all disciplines because equilibrium systems are themselves something of an idealization and met only infrequently (in the strictest sense) in real life.

Time-variable-wise, the first problem to address is pretty clear. Chapter 3 used composite systems to illustrate fluctuations about maximum entropy states. It was shown that a pressure measurement is preceded by uncertainty on the chemist's part.

This determined the amount of information trapped by a barometer. Yet here is precisely where the time issues need to be drawn out. If the chemist endeavors a second measurement, his or her uncertainty will depend on how much time has elapsed. If the system is allowed insufficient time to explore the possible configurations, there is less uncertainty and thus less information to purchase. If the second measurement is performed after a long time, the system has had time to forget where it was. The original information condition is restored.

In short, the thermodynamic information of a system, even at equilibrium, is neither static nor uniform. The principle to illuminate would be the time-dependence, and the related effects of system size and composition. The results would be not simply $I_{X \leftrightarrow p}$, $I_{X \leftrightarrow V}$, and $MI_{XY \leftrightarrow pV}$, but rather $I_{X \leftrightarrow p}(t)$, $I_{X \leftrightarrow V}(t)$, and $MI_{XY \leftrightarrow pV}(t)$. To arrive at these quantities, the transport properties of the system would have to be charted. As is well appreciated, systems demonstrate relaxation times τ that are characteristic of the gradient of interest—pressure, temperature, and chemical. The chemist's second measurement of p, T, C_p, and so forth at $t < \tau$ affords less information than at $t > \tau$. Along the same lines, all fluctuations are not alike. If a barometer registers p at $t = 0$ as $<p> + 1.50\sigma_p$, the uncertainty at a later time will be different from the case where $p(t=0) = <p> - 2.50\sigma_p$.

Issues surrounding time impact more than just the system. A barometer must establish equilibrium to communicate the correct number of pascals. A thermometer must establish thermal equilibrium for high fidelity readings. If the allotted time is too short in either case, then errors will plague the information purchase. Mechanical and thermal waves do not propagate at the same rate and phase. Thus the errors regarding temperature will differ from those of pressure. It must also be noted that a measurement of T generally affects that of p and vice versa. The interference effects determine the limits to which thermodynamic information of different variables can be processed in parallel.

Chapters 4 and 5 examined transformations effected by variable tuning and energy exchanges. Time was irrelevant throughout since the pathways were all taken as reversible. Time issues require consideration here as well, however. If the time allotted for program execution is insufficient, then irreversibilities will transpire in the system as side effects. The state points will be relocated imperfectly since each step will reflect the system history plus mechanical wear and tear. Variables such as p, V, and n specify the entire state point locus for a reversible transformation. Matters are much more complicated if t must be included. As stated at the beginning of Chapter 1, the word *information* motivates much discussion.

8.2 SOURCES AND FURTHER READING

Equilibrium thermodynamics has been thoroughly charted over decades; so has nonequilibrium thermodynamics. The author's shelf includes excellent texts by Yourgrau, van der Merwe, and Raw [1], Haase [2], and de Groot and Mazur [3]. Desloge's book addresses several nonequilibrium fundamentals [4]; the same is true for Callen's [5]. The books by Morowitz present significant insights about nonequilibrium states [6,7]. Stanley's text [8] and Denbigh's [9] have educated the author regarding phase transitions and phase rule applications. The book by Berry, Kazakov, Tsirlin, Sieniutycz, and Szwast is a must-read regarding thermodynamics with time included as a variable [10]. There is much to learn from this source about the horizon and beyond. The short paper by Schreiber notably impacted the author's appreciation of information, time, and their interface [11]. The same statement applies to Mackey's treatise on the arrow of time [12].

8.3   SUGGESTED EXERCISES

Chapter 1 introduced the topics of the book qualitatively. The succeeding chapters examined them in quantitative terms. Hopefully, the reader's perspective is enhanced to some degree, having gone through the text. To consider the changes, three of the Chapter 1 exercises are urged for repetition. Compare one's answers with those written the first time around.

8.1   Chapter 1 opened with the statement, "Information motivates much discussion." Several declarations followed. Choose one and write a two- to three-page response paper. The response should argue the merits and deficiencies regarding information.

8.2   Chapter 1 presented the idea that information represents a system's capacity for controlling work and heat transactions. As in the first exercise, compose a response paper that addresses the merits and deficiencies of the idea.

8.3   Describe two examples drawn from chemistry where probability plays a role. Do likewise regarding conditional probability.

REFERENCES

[1] Yourgrau, W., van der Merwe, A., Raw, G. 1982. *Treatise on Irreversible and Statistical Thermophysics*, Dover, New York.
[2] Haase, R. 1990. *Thermodynamics of Irreversible Processes*, Dover, New York.
[3] de Groot, S. R., Mazur, P. 1984. *Non-Equilibrium Thermodynamics*, Dover, New York.
[4] Desloge, E. A. 1968. *Thermal Physics*, Holt, Rinehart, and Winston, New York.
[5] Callen, H. B. 1960. *Thermodynamics: An Introduction to the Physical Theories of Equilibrium Thermostatics and Irreversible Thermodynamics*, Wiley, New York.
[6] Morowitz, H. J. 1970. *Energy for Biologists: An Introduction to Thermodynamics*, Academic Press, New York.
[7] Morowitz, H. J. 1979. *Energy Flow in Biology: Biological Organization as a Problem in Thermal Physics*, Ox Bow Press, Woodbridge, CT.
[8] Stanley, H. E. 1971. *Introduction to Phase Transitions and Critical Phenomena*, Oxford University Press, New York.
[9] Denbigh, K. 1973. *The Principles of Chemical Equilibrium,* Cambridge University Press, Cambridge.
[10] Berry, R. S., Kazakov, V., Tsirlin, A. M., Sieniutycz, S., Szwast, Z. 2000. *Thermodynamic Optimization of Finite Time Processes*, Wiley, New York.
[11] Schreiber, T. 2000. Measuring Information Transfer, *Phys. Rev. Lett.* 85, 461.
[12] Mackey, M. C. 2003. *Time's Arrow*, Dover, New York.

Appendix A: Source Program for Constructing Molecular Message Tapes and Computing Information

This program can be adapted readily to small organic compounds by changing the atom vector and bond matrix components in subroutine 1000. As written, the program addresses the information properties of the following alcohol derivative of cyclohexene:

Statements useful for verification and troubleshooting have been disengaged by REM prefixes; these can be re-activated as needed. BASIC has been used as the source code for simplicity along with the resident random number generator. The algorithm is readily adapted to higher order analyses, longer record tapes, and to C, C++, and Pascal codes.

```
1 REM REM SET UP FOR AN ALCOHOL DERIVATIVE OF CYCLOHEXENE
10 INPUT "R EQUALS????", R: RANDOMIZE R: REM SEEDS
RANDOM NUMBER GENERATOR
20 DIM A%(17), B%(17, 17), M$(500), SUM(500)
30 D% = 17: M% = 7: REM MATRIX DIMENSION AND NUMBER OF
MAJOR ATOMS
40 GOSUB 1000: REM SET UP ATOM ARRAY AND BOND MATRIX
44 W3% = 45: REM SET INITIAL W3% TO ANY INTEGER >
MATRIX DIMENSION
45 FOR W5% = 1 TO 5000: REM SETS WALK SIZE
50 IF W5% = 1 THEN S% = INT(RND * M% + 1): REM SELECT
FIRST MAJOR ATOM
52 REM PRINT "S% EQUALS.....", S%: REM INPUT
"CONTINUE????", ANS1
60 B% = 0
70 FOR K% = 1 TO D%
80 IF B%(S%, K%) > 0 THEN B% = B% + 1: REM COUNT
COVALENT BOND LINKS
90 NEXT K%
100 B2% = INT(RND * B% + 1): REM PICK RANDOM LINK
```

```
105 REM PRINT "B%, B2% EQUALS....", B%, B2%: REM INPUT
"CONTINUE????", ANS1
120 B3% = 0: REM GET READY TO COUNT AGAIN
125 W1% = S%: REM TAG ATOM 1
130 FOR K% = 1 TO D%
140 IF B%(S%, K%) > 0 THEN B3% = B3% + 1
145 IF B3% = B2% THEN W2% = K% ELSE 150
146 REM PRINT „W2% EQUALS....", W2%: REM INPUT
"CONTINUE????", ANS1
147 IF W2% = W3% THEN 100 ELSE 160
150 NEXT K%
160 REM WE HAVE A STATE, NOW CHECK
170 GOSUB 2000
180 IF W5% = 1 THEN S2$ = S$ ELSE S2$ = S2$ + S$
190 REM IF W5% MOD 200 = 0 THEN PRINT S2$: INPUT
"CONTINUE?????", ANS1
200 IF A%(W2%) = 6 OR A%(W2%) = 8 THEN S% = W2%: W3% = W1%
210 IF A%(W2%) = 1 THEN S% = W1%: W3% = W2%
300 NEXT W5%
302 PRINT S2$
310 GOSUB 3000: REM PARSE RECORD TAPE
320 GOSUB 5000: REM FIGURE MI
900 END
1000 REM SUB TO SET UP ATOM ARRAY AND BOND MATRIX
1005 REM FIRST ZERO THINGS
1010 FOR J% = 1 TO D%
1020 A%(J%) = 0
1030 NEXT J%
1040 FOR J% = 1 TO D%
1050 FOR K% = 1 TO D%
1060 B%(J%, K%) = 0
1070 NEXT K%
1080 NEXT J%
1100 FOR J% = 1 TO D%
1110 IF J% <= 6 THEN A%(J%) = 6
1112 IF J% = 7 THEN A%(J%) = 8
1114 IF J% > 7 THEN A%(J%) = 1
1120 NEXT J%
1200 B%(1, 2) = 1
1210 B%(1, 6) = 1
1220 B%(2, 3) = 2
1230 B%(3, 4) = 1
1240 B%(4, 5) = 1
1250 B%(5, 6) = 1
1260 B%(6, 1) = 1
```

```
1262 B%(6, 7) = 1
1270 REM NOW HELP COMPLETE BOND MATRIX
1280 FOR J% = 1 TO D%
1290 FOR K% = 1 TO D%
1300 IF B%(J%, K%) > 0 AND B%(K%, J%) = 0 THEN B%(K%,
J%) = B%(J%, K%)
1310 NEXT K%
1320 NEXT J%
1330 M2% = M%: REM MAJOR ATOMS
1400 FOR J% = 1 TO M%
1410 B% = 0
1420 FOR K% = 1 TO M%
1430 IF B%(J%, K%) > 0 THEN B% = B% + B%(J%, K%)
1440 NEXT K%
1450 IF A%(J%) = 6 THEN V% = 4 - B%: REM VALENCE RULE
1451 IF A%(J%) = 8 THEN V% = 2 - B%: REM VALENCE RULE
1452 PRINT J%, B%, V%, M2%: REM INPUT "CONTINUE?????",
ANS1
1460 FOR K% = (M2% + 1) TO (M2% + V%)
1470 B%(J%, K%) = 1: B%(K%, J%) = 1: REM H-BOND
1480 NEXT K%
1490 M2% = M2% + V%
1500 NEXT J%
1510 REM NOW READ BACK
1520 FOR J% = 1 TO D%
1530 FOR K% = 1 TO D%
1540 IF B%(J%, K%) > 0 THEN PRINT J%, K%, B%(J%, K%):
REM INPUT «CONTINUE????», ANS1
1550 NEXT K%
1560 NEXT J%
1900 RETURN
2000 REM SUB TO CHECK STATES
2005 REM C-H = "A"; C-C = "B"; C=C = "C"; C-O = "D"; O-H
= "E"
2006 REM ADD STATES TO THIS SUBROUTINE AS NEEDED
2010 IF A%(W1%) = 6 AND A%(W2%) = 1 THEN S$ = "A": GOTO
2100
2020 IF A%(W1%) = 6 AND A%(W2%) = 6 AND B%(W1%, W2%) = 1
THEN S$ = "B": GOTO 2100
2030 IF A%(W1%) = 6 AND A%(W2%) = 6 AND B%(W1%, W2%) = 2
THEN S$ = "C": GOTO 2100
2032 IF A%(W1%) = 6 AND A%(W2%) = 8 AND B%(W1%, W2%) = 1
THEN S$ = "D": GOTO 2100
2034 IF A%(W1%) = 8 AND A%(W2%) = 6 AND B%(W1%, W2%) = 1
THEN S$ = "D": GOTO 2100
```

```
2036 IF A%(W1%) = 8 AND A%(W2%) = 1 AND B%(W1%, W2%) = 1
THEN S$ = "E": GOTO 2100
2100 RETURN
3000 REM SUB TO PARSE RECORD TAPE
3005 L% = LEN(S2$)
3006 FOR M% = 1 TO 4: REM ORDERS
3007 C% = 0: N = 0: H1 = 0: H2 = 0: REM INITIATE STATE
COUNTER
3008 FOR W2% = 1 TO 500: SUM(W2%) = 0: NEXT W2%: REM
ZERO STATE ARRAY
3009 H = 0
3010 FOR J% = 1 TO L% - M% + 1
3012 A$ = MID$(S2$, J%, M%)
3013 H = 0
3014 FOR W2% = J% TO J% + M% - 1
3016 B$ = MID$(S2$, W2%, 1): GOSUB 4000: REM GET ABA
ENERGY
3018 H = H + H5
3019 NEXT W2%
3020 FOR K% = 1 TO C%
3030 IF A$ = M$(K%) THEN SUM(K%) = SUM(K%) + 1: N = N +
1: GOTO 3050: REM OLD STATE
3040 NEXT K%
3042 REM WE HAVE NEW STATE
3045 C% = C% + 1
3046 M$(C%) = A$: SUM(C%) = 1: N = N + 1
3050 H1 = H1 + H: H2 = H2 + H * H
3055 NEXT J%
3060 REM NOW FIGURE SHANNON INFORMATION, ABA ENERGY
AND DISPERSION
3070 INFO = 0
3080 FOR K% = 1 TO C%
3090 INFO = INFO - (SUM(K%) / N) * LOG(SUM(K%) / N)
3100 NEXT K%
3110 PRINT "ORDER, INFO EQUALS.....", M%, INFO / LOG(2)
3120 H1 = H1 / N: H2 = (H2 / N) - H1 * H1
3130 PRINT "ABA AVG ENERGY, SIG EQUAL....", H1, SQR(H2):
INPUT "CONTINUE????", ANS1
3200 NEXT M%
4000 REM SUB TO ASSIGN BOND ENERGIES
4005 REM ADD STATES AND ABA ENERGIES TO THIS SUBROUTINE
AS NEEDED
4010 IF B$ = "A" THEN H5 = 414: REM kilojoules/mole
4020 IF B$ = "B" THEN H5 = 347
4030 IF B$ = "C" THEN H5 = 612
4032 IF B$ = "D" THEN H5 = 351
```

```
4034 IF B$ = "E" THEN H5 = 464
4200 RETURN
5000 REM SUB TO FIGURE MI
5001 PRINT : PRINT "NOW FIGURING MUTUAL INFORMATION...."
5006 FOR M% = 2 TO 4: REM ORDERS
5007 C% = 0: N = 0: REM INITIATE STATE COUNTER
5008 FOR W2% = 1 TO 500: SUM(W2%) = 0: NEXT W2%: REM
ZERO STATE ARRAY
5010 FOR J% = 1 TO L% - M% + 1
5012 A$ = MID$(S2$, J%, M%)
5020 FOR K% = 1 TO C%
5030 IF A$ = M$(K%) THEN SUM(K%) = SUM(K%) + 1: N = N +
1: GOTO 5050: REM OLD STATE
5040 NEXT K%
5042 REM WE HAVE NEW STATE
5045 C% = C% + 1
5046 M$(C%) = A$: SUM(C%) = 1: N = N + 1
5050 REM
5055 NEXT J%
5060 REM NOW FIGURE MI
5070 MI = 0
5080 FOR K% = 1 TO C%
5081 APPLE1 = SUM(K%) / N: APPLE2 = APPLE1
5082 FOR W1% = 1 TO M%
5084 B$ = MID$(M$(K%), W1%, 1)
5085 N5 = 0
5086 FOR W2% = 1 TO L%
5088 IF MID$(S2$, W2%, 1) = B$ THEN N5 = N5 + 1
5090 NEXT W2%
5092 APPLE2 = APPLE2 / (N5 / L%)
5094 NEXT W1%
5096 MI = MI + APPLE1 * LOG(APPLE2)
5100 NEXT K%
5110 PRINT "ORDER, MI EQUALS.....", M%, MI / LOG(2)
5200 NEXT M%
5500 RETURN
```

Appendix B: Answers to Selected Exercises

CHAPTER 1

1.4 The sum of fractions along the ordinate scales as the square root of the sum of fractions along the abscissa.

1.7 There are 80 allowed combinations.

CHAPTER 2

2.1 (a) 8.50 Hartleys = 28.2 bits (b) 5.30 nits = 2.30 Hartleys

2.5 The sum of weighted surprisals need not mirror the distribution function since the indexing of states is arbitrary.

2.6 (a) c. 10^{90} peptides (b) 300 bits

2.9 (a) 0.286 bits (c) 0.406 bits (d) 0.583 bits (e) 0.0523 bits (f) 4.89×10^{-4} bits

2.11 (a) $c. 10^{103}$ isomers (b) 1.47 bits

2.12 (a) 5.14×10^{-3} bits (b) 7.11×10^{-6} bits (c) 1.27×10^{-8} bits

CHAPTER 3

3.8 (a) $\beta_T^{vdW} = -\frac{1}{V}\left(\frac{\partial V}{\partial p}\right)_{T,n} = \frac{V^2 - nbV}{3pV^2 - 2pnbV + an^2 - 2nRT}$

$\beta_T^{Diet} = -\frac{1}{V}\left(\frac{\partial V}{\partial p}\right)_{T,n} = \frac{nb - V}{\left(\dfrac{an}{V}\right)\cdot\exp\left[\dfrac{-a}{RTV}\right] - pV}$

3.10 The Legendre transform cancels to zero.

3.14 For $\lambda = 0.05$, 0.10, and 1.00, the respective sector volumes are approximately 0.0379, 0.0100, and 10^{-4} meters^3.

3.15 (a) $\langle p \rangle \approx 1.58 \times 10^6$ pascals (b) $\sigma_p \approx 6.12 \times 10^{-6}$ pascals

CHAPTER 4

4.1 (a) $\Lambda_{min} = 1$
4.2 (c) $I_{X \leftrightarrow p} \approx 7.64$ bits (d) $I_{X \leftrightarrow T} \approx 6.96$ bits
4.3 (a) $\sigma_p^{(A)} \approx 683$ pascals; $\sigma_p^{(B)} \approx 563$ pascals
4.4 (c) $\langle p \rangle \approx 6.9 \times 10^5$ pascals, $\sigma_p \approx 2.4 \times 10^5$ pascals (d) $I_{X \leftrightarrow p} \approx 5.4$ bits
4.5 (a) $\langle p \rangle \approx 5.4 \times 10^5$ pascals, $\sigma_p \approx 2.3 \times 10^5$ pascals (b) $\varepsilon \approx 0.12$
4.6 Only one adiabat can intersect a given state.
4.10 $MI_{XY \leftrightarrow pV} \approx 2.93$ bits

CHAPTER 5

5.2 The less exacting chemist is not overly disadvantaged. The average yield is $c.$ 0.60. The standard deviation is $c.$ 0.0034.
5.5 (a) $\Omega_{X \leftrightarrow p} \approx 1.18$.
5.7 The isotherm offers a more favorable $\Omega_{X \leftrightarrow p}$.
5.8 $\langle \Omega_{X \leftrightarrow p} \rangle \approx 65$; $\sigma_{\Omega_{X \leftrightarrow p}} \approx 11$. The distribution is fairly uniform.
5.9 (a) **E** expresses smaller $I_{X \leftrightarrow U}$.

CHAPTER 6

6.2 $\xi_{MI}^{cyclobut} = 0.0971$, $\sigma_{\xi,MI}^{cyclobut} = 3.9 \times 10^{-3}$ bits/message unit

$\xi_{MI}^{cubane} = 0.122$, $\sigma_{\xi,MI}^{cubane} = 5.1 \times 10^{-4}$ bits/message unit

$\xi_{\sigma_D}^{cyclobut} = 5.40$, $\sigma_{\xi,\sigma_D}^{cyclobut} = 0.17$ kilojoules/mole · message unit

$\xi_{\sigma_D}^{cubane} = 4.71$, $\sigma_{\xi,\sigma_D}^{cubane} = 0.15$ kilojoules/mole · message unit

6.3 $\xi_{MI}^{cyclobutene} = 0.336$, $\sigma_{\xi,MI}^{cyclobutene} = 2.7 \times 10^{-2}$ bits/message unit

$\xi_{MI}^{Dewar\ Benzene} = 0.358$, $\sigma_{\xi,MI}^{Dewar\ Benzene} = 2.3 \times 10^{-2}$ bits/message unit

$\xi_{\sigma_D}^{cyclobutene} = 18.6$, $\sigma_{\xi,\sigma_D}^{cyclobutene} = 0.99$ kilojoules/mole · message unit

$\xi_{\sigma_D}^{Dewar\ Benzene} = 19.6$, $\sigma_{\xi,\sigma_D}^{Dewar\ Benzene} = 1.1$ kilojoules/mole · message unit

CHAPTER 7

7.1 (a) 451 joules/mole (b) 1.57 bits
7.2 (a) –4700 joules/mole (b) –4040 joules/mole
7.11 Molecules that are optical isomers have identical standard chemical potentials and molar enthalpies in the gas phase.

Index

Printed and bound by CPI Group (UK) Ltd, Croydon, CR0 4YY

21/10/2024

01777089-0006